4th Symposium on Large TPCS for Low Energy Rare Event Detection 2008

Journal of Physics: Conference Series Volume 179

Paris, France
18-19 December 2008

ISBN: 978-1-61738-305-2

Printed from e-media with permission by:

Curran Associates, Inc.
57 Morehouse Lane
Red Hook, NY 12571
www.proceedings.com

Some format issues inherent in the e-media version may also appear in this print version.

Printed by Curran Associates, Inc. (2010)

Institute of Physics
Dirac House, Temple Back
Bristol BS1 6BE UK

Phone: 44 1 17 929 7481
Fax: 44 1 17 920 0979

techtracking@iop.org

TABLE OF CONTENTS

Physics and Cosmology: The Milli-electron-volt Scale 1
Eduard Massó

The Test Beams of the T2K TPC 5
Claudio Giganti

Spherical TPC Development and Trends 11
S Andriamonje, S Aune, E Bougamont, M Chapellier, P Colas, J Derré, E Ferrer-Ribas, S Gaffet, G Gerbier, I Giomataris, M Gros, I G Irastorza, P Mangier, R Lemrani, X F Navick, P Piquemal, P Salin, I Savvidis, J D Vergados, M Zampaolo

Double Beta Decay in Liquid Xenon 16
Razvan Gornea

The NEXT Experiment 22
J Díaz, N Yahlali, M Ball, J A S Barata, F I G M Borges, E Calvo, S Cárcel, J M Carmona, S Cebrián, A Cervera, X Cid, C A N Conde, T Dafni, T H V T Dias, L M P Fernandes, E Ferrer-Ribas, E D C Freitas, J Galán, A Gil, I Gil, I Giomataris, H Gómez, J J Gómez-Cadenas, F Granena, J A Hernando-Morata, F J Iguaz, I Irastorza, J A M Lopes, D Martínez, C M B Monteiro, J Muñoz-Vidal, C Palomares, I Irastorza, M Lázaro, T Lux, G Luzón, J Martín-Albo, F Monrabal, J Morales, F Nova, P Novella, D Nygren, L Ripoll, A Rodríguez, J Ruz, N F Sánchez, F P Santos, J M F Dos Santos, L Serra, M Sorel, L M N Távora, A Tomás, J F C A Veloso, A Verdugo

Topological Detection of Double Beta Decay with NEMO3 and SuperNEMO 30
Ruben Saakyan

Energy Resolution of Alpha Particles in a Micromegas Detector at High Pressure 37
F J Iguaz, T Dafni, E Ferrer-Ribas, G Giomataris, P Gordetzsky, I Irastorza, P Salin, A Tomás

New Results of GridPix TPCs 43
Harry Van Der Graaf

The DMTPC Project 46
Gabriella Sciolla

Directional Neutron Detection and TPC Development at LLNL 54
M Heffner

Micromegas μTPC for Direct Dark Matter Search with MIMAC 56
F Mayet, O Guillaudin, C Grignon, C Koumeir, D Santos, P Colas, I Giomataris

Low Energy Measurements with Helium Micromegas μTPC 62
O Guillaudin, F Mayet, C Grignon, C Koumeir, D Santos, P Colas, I Giomataris

The Gotthard and MUNU TPCs: From the Past to the Future? 67
Jean-Luc Vuilleumier

Optical Readout of Secondary Scintillation from Liquid Argon Generated by a Thick Gas Electron Multiplier 72
P K Lightfoot, G J Barker, K Mavrokoridis, Y A Ramachers, N J C Spooner

An Ultra-low-background Detector for Axion Searches 85
S Aune, T Dafni, G Fanourakis, E Ferrer Ribas, J Galán Lacarra, T Geralis, I Giomataris, F J Iguaz, I G Irastorza, K Kousouris, J Morales, J P Mols, T Papaevangelou, J Ruz, A Tomás, T Vafeiadis

Relic Neutrino Detection Using Neutrino Capture on Beta Decaying Nuclei 91
Alfredo G Cocco

Author Index

The Fourth International Symposium on large TPCs for low-energy rare-event detection was held at the Hermite auditorium of the Insitute Henri Poincaréte;, 11 rue Pierre et Marie Curie in Paris on 18–19 December 2008. As in previous instances of the meeting, held always in Paris in 2006, 2004 and 2002, it gathered a significant community of physicists involved in rare event searches and/or development of time projection chambers (TPCs).

The purpose of the meeting was to present and discuss the status of current experiments or projects involving the use of large TPCs for the search of rare events, like low-energy neutrinos, double beta decay, dark matter or axion experiments, as well as to discuss new results and ideas in the framework of the last developments of Micro Pattern Gaseous Detectors (MPGD), and how these are being – or could be – applied to the mentioned searches. The rapid evolvement of these devices and the relevance of their latest results need to be efficiently transferred to the rare event community. The creation of this series of meetings followed the motivation of bringing together both know-hows and it is proving to be a fruitful area of collaboration.

Once more, the format of the meeting proved to be a success. A short (2 days) and relatively informal program with some recent highlighted results, rather than exhaustive reviews, attracted the interest of the audience. The symposium, fourth of the series, is becoming consolidated as a regular meeting place for the synergic interplay between the fields of rare events and TPC development.

Apart from the usual topics central to the conference subject, like the status of some low-energy neutrino physics and double beta decay experiments, dark matter experiments (and in general physics in underground laboratories), axion searches, or development results, every year the conference programme is enriched with original slightly off-topic contributions that trigger the curiosity and stimulate further thought. This time we enjoyed some talks on cosmology and the possible detection of relic neutrinos.

The present volume contains the proceedings of the conference. All papers published here have been peer reviewed through process administered by the Editors to meet standards for publication. They correspond to 16 of the talks given in the conference and they all represent high quality work in the above mentioned fields.

The successful organization of the Symposium was made possible thanks to the contribution of the members of the Organizing Committee and International Advisory Committee. We want to thank in particular the session chairmen, J Iliopoulos, Ph Gorodetzky, B Mansoulié, J-E Augustin, D Santos, S Jullian and S Loucatos for their contribution to the smooth running of the workshop. The symposium was free of fees, which was made possible thanks to the financial support from CEA/IRFU and CNRS/IN2P3, the two major French research organizations that are gratefully acknowledged. We greatly acknowledge the contribution of the University of Zaragoza to the publication of these proceedings.

Finally we want to thank the speakers for the high quality of their talks and all participants for coming to Paris and actively contributing in the meeting.

All papers published in this volume of *Journal of Physics: Conference Series* have been peer reviewed through processes administered by the proceedings Editors. Reviews were conducted by expert referees to the professional and scientific standards expected of a proceedings journal published by IOP Publishing.

Journal of Physics: Conference Series **179** (2009) 012001 doi:10.1088/1742-6596/179/1/012001

Physics and Cosmology : the milli-electron-Volt Scale

Eduard Massó
Grup de Física Teòrica and Institut de Física d'Altes Energies
Universitat Autònoma de Barcelona, 08193 Bellaterra, Spain

Abstract. A short review about vacuum energy and the cosmological constant is presented. The observed acceleration of the universe introduces a new meV energy scale. The problem is that, theoretically, the predicted vacuum energy is many orders of magnitude larger than 10^{-3} eV. The problem is a link between two Standard Models, namely the Standard Model of Particles and their Interactions (where the vacuum energy appears) and the Standard Cosmological Model (where a cosmological constant is a good fit to data), and perhaps it is a clue in our search for new physics.

1. The Cosmological Constant

Let us mention some important facts about our universe. First of all, we know our universe is expanding because of the measured redshift in the spectral lines of light sent by distant objects. Consider the moment of emission of a photon from a galaxy which is at a (physical) distance d from the Earth. The photon is emitted with a (proper) wavelength λ_e. At the moment of reception of the photon at the Earth the physical distance between the galaxy and us has increased to d_0 and the wavelength λ_r of the photon has increased in the same proportion

$$\frac{\lambda_r}{\lambda_e} = \frac{d_0}{d} = 1 + z \tag{1}$$

where z is the redshift.

Introducing the assumption of homogeneity and isotropy, we can describe our universe quite simply. Before showing it, we introduce the concept of comoving distance and scale factor. Take the example of the galaxy we introduced before and the Earth. Their comoving distance x is fixed and tells us which two objects we are considering. The expansion can be described in terms of a time dependent scale factor $a(t)$, in such a way that at the moment of emission the scale factor has a value $a = d/x$ while at reception it is $a_0 = d_0/x$.

The equations governing the behavior of the scale factor can be deduced from the Einstein equations of General Relativity,

$$R_{\mu\nu} - \frac{1}{2} g_{\mu\nu} R + \Lambda\, g_{\mu\nu} = 8\pi G_N\, T_{\mu\nu} \tag{2}$$

where $T_{\mu\nu}$ is the energy-momentum tensor of the universe content and the parameter Λ is the (in)famous cosmological constant. Let us first set $\Lambda = 0$. In this case, from the Einstein equations we obtain the Friedmann equations

$$H^2 = \left(\frac{\dot{a}}{a}\right)^2 = \frac{8\pi G_N}{3}\,\rho \tag{3}$$

$$\frac{\ddot{a}}{a} = -\frac{4\pi G_N}{3}(\rho + 3p) \tag{4}$$

Here ρ and p include the energy density and pressure of the different fluids that form the universe. We have simplified a bit the equations because we have written them in the case of a flat universe, which is a good approximation.

Observations at recent z favor a universe with $\ddot{a} > 0$, i.e., an accelerating universe. We know from independent observations that there is a large density of cold dark matter in the universe. However its contribution to (4) would lead to $\ddot{a} < 0$ since $\rho_m > 0$ and $p_m = 0$ ("m" means both visible matter and dark matter). Actually, fits to the data lead to p in (4) to be negative and on the order of magnitude of $p \sim -\rho$. From where we may have negative pressure contributions?

The cosmological constant may come to our rescue. The full Friedmann equations when Λ is present are

$$H^2 = \left(\frac{\dot{a}}{a}\right)^2 = \frac{8\pi G_N}{3}\rho + \frac{\Lambda}{3} \tag{5}$$

$$\frac{\ddot{a}}{a} = -\frac{4\pi G_N}{3}(\rho + 3p) + \frac{\Lambda}{3} \tag{6}$$

We now see that the Λ contribution is able to give an accelerating universe, $\ddot{a} > 0$.

It is convenient to write an equivalent density ρ_Λ and pressure p_Λ for the cosmological constant contribution,

$$\frac{\Lambda}{3} \equiv \frac{4\pi G_N}{3}\rho_\Lambda \tag{7}$$

$$p_\Lambda = -\rho_\Lambda \tag{8}$$

With these definitions the form of the Friedmann equations is simply given by (3) and (4), but now understanding that there are not only from visible matter and dark matter but also from Λ,

$$\begin{aligned} \rho &= \rho_m + \rho_\Lambda + \ldots \\ p &= p_m + p_\Lambda + \ldots \end{aligned} \tag{9}$$

where the dots are potential components of the universe apart from matter and Λ. (We neglect radiation because it does not play any role in the Friedmann equations in the universe at recent z).

It is worth noticing that these two components have a very simple equation of state relating density and pressure,

$$p_i = w_i \rho_i \tag{10}$$

with w_i constant. Indeed, we have

$$\begin{aligned} w_m &= 0 \\ w_\Lambda &= -1 \end{aligned} \tag{11}$$

In these cases one can integrate the equation

$$\frac{d}{dt}(\rho_i a^3) + p_i \frac{d}{dt}(a^3) = 0 \tag{12}$$

which can be deduced from (3) and (4), and which expresses energy conservation. One gets for matter

$$\rho_m \propto \frac{1}{a^3} \tag{13}$$

expressing the fact that volume goes as a^3 in an expanding universe. However, for the cosmological constant one gets

$$\rho_\Lambda \propto \text{constant} \tag{14}$$

2. Dark Energy

To obtain the values for the matter and the cosmological constant densities, one uses data from supernovae, cosmic microwave background, cluster analysis, large scale structure, age, lensing, etc. What one gets is a consistent and robust picture, called the Standard Cosmological Model where the total density of universe is very near the critical density ρ_c (defined as the density that would make the universe exactly flat),

$$\rho_m + \rho_\Lambda \simeq \rho_c \tag{15}$$

and where the contributions are

$$\Omega_m = \frac{\rho_m}{\rho_c} \simeq 0.3 \tag{16}$$

and

$$\Omega_\Lambda = \frac{\rho_\Lambda}{\rho_c} \simeq 0.7 \tag{17}$$

with small errors.

This last contribution can be expressed back in terms of ρ_Λ, and one gets

$$\rho_\Lambda \simeq (2 \times 10^{-3}\ \mathrm{eV})^4 \tag{18}$$

This is the new meV energy scale that characterizes the cosmological contribution. To understand its origin is a challenge for contemporary physics.

We have used the contribution of the cosmological constant to describe the acceleration of the universe, but it is time now to make some generalization. We could have contributions in (4) that lead to acceleration without the need to be exactly of the form of a cosmological constant. We could have for instance $w \neq -1$, but still leading to acceleration. We see from (4) that $w < -1/3$ gives $\ddot{a} > 0$. In all generality we could have a z-dependent $w = w(z)$. We call dark energy any component of the universe leading to acceleration. We have in this case a density ρ_{de} that today is measured as

$$\rho_{de} \simeq (2 \times 10^{-3}\ \mathrm{eV})^4 \tag{19}$$

A popular model of dark energy is quintessence, where fields which are slowly varying in time are the origin of the acceleration.

3. Vacuum Energy

The vacuum energy acts as a cosmological constant. This is due to the fact that general covariance gives the form of the expectation value in the vacuum of the energy-momentum tensor,

$$T^{\mu\nu}_{vac} = <0|T^{\mu\nu}|0> = \rho_{vac}\ g^{\mu\nu} \tag{20}$$

This might be good news, because vacuum energy could be an explanation for having a cosmological constant Λ.

The problem arises when we realize that the Standard Model of Particles is a Quantum Field Theory, and that in vacuum the expectation value of the Hamiltonian does not vanish because it receives contributions of all zero-point energies of all modes for all fields

$$<0|H|0> = \sum_{fields} \eta_i \int_0^\infty \frac{d^3k}{(2\pi)^3} \frac{1}{2}\hbar\sqrt{k^2 + m_i^2} \tag{21}$$

Here $\eta_{boson} = 1$ and $\eta_{fermion} = -1$. The integral is divergent so that we obtain infinite as the result.

What we do in Quantum Field Theory is to subtract this (infinite) quantity,

$$H \longrightarrow H - <0|H|0> \tag{22}$$

or, in more technical terms, we use normal ordering. The argument is that we are only concerned with energy differences, so that the subtraction does not have any physical consequences. However, this is true when we do not consider gravitational effects and it is no longer true when gravitation is on.

The divergence in (21) comes because we integrate over all energies and momenta, or, in other words, we believe the theory is valid up to infinite energies. Let us be more modest and assume we can use the Quantum Field theory until a maximum cutoff k_{max}. Thus for a single (bosonic) field we have

$$\rho_{vac} = \frac{1}{4\pi^2} \int_0^{k_{max}} dk\ k^2 \sqrt{k^2 + m^2} = \frac{k_{max}^4}{16\pi^2} \tag{23}$$

(we use $m \ll k_{max}$).

If we believe that our theories can be used until the Planck mass, then we get a total vacuum energy

$$\rho_{vac} = \frac{M_P^4}{16\pi^2} \simeq (10^{18}\ \mathrm{GeV})^4 \tag{24}$$

We see that this estimate exceeds by many orders of magnitude the measured value (19). Perhaps we are still too pretentious because in reality we have probed the Standard Model of Particles only until the Fermi scale M_W. If we use this cutoff we get

$$\rho_{vac} = \frac{M_W^4}{16\pi^2} \simeq (10\ \mathrm{GeV})^4 \tag{25}$$

still 13 orders of magnitude in energy scale.

In conclusion, vacuum energy contributes to the energy density of the universe many orders of magnitude above observation. Even if we find a theory that explains the acceleration of the universe with the correct value (19), we should find a satisfactory explanation to the problem of the vacuum energy contribution to the cosmological constant. Perhaps when trying to solve this challenging problem we will get hints leading to the discovery of new physics.

The test beams of the T2K TPC

Claudio Giganti for the T2K/ND280/TPC Collaboration
DSM/Irfu/SPP, CEA Saclay, F-91191 Gif Sur Yvette, France
E-mail: claudio.giganti@cea.fr

Abstract. T2K is a long baseline neutrino experiment that will search for $\nu_\mu \rightarrow \nu_e$ oscillation. A near detector complex (ND280) will be installed at 280 meters from the neutrino production point and will be used to measure the neutrino energy spectrum, flavor content and neutrino cross-sections of the unoscillated neutrino beam. ND280 includes three large TPCs based on bulk-MicroMegas technology as readout device. Performances of bulk-MicroMegas modules with an $Ar/iC_4H_{10}/CF_4$ (95/2/3) gas mixture are presented. First results of beam tests performed with the first T2K/TPC are also discussed and in particular measurements of the deposited energy resolution and particle identification capability of the TPC will be presented.

1. The T2K experiment

The T2K (Tokai to Kamioka) experiment is a long baseline neutrino oscillation experiment mainly designed to search for $\nu_\mu \rightarrow \nu_e$ oscillation[1]. The Japan Proton Accelerator Research Complex (JPARC) at Tokai (Japan) will provide an high intensity proton beam, accelerating protons up to 30 GeV. These protons interacting with a graphite target will produce mesons that will then decay into neutrino in a decay volume. In this way a ν_μ beam is produced.
The neutrino will travel to SuperKamiokande[2], a water Cerenkov detector at a distance of 295 km from the neutrino production point. The main physics goals of the T2K experiment are the precise measurement of the parameters Δm^2_{23} and θ_{23} and the first measurement of θ_{13}, the only angle not yet measured in the neutrino mixing matrix. The existing upper limit for this angle comes from the CHOOZ reactor experiment[3] that performed the measurement in the ν_e disappearance channel. The T2K experiment instead open the possibility to measure this angle in an appearance experiment. The main background to this measurement comes from neutral current π^0 production and from the intrinsic ν_e contamination in the beam.

For the T2K physics goals it is not sufficient to measure neutrinos only in the far detector, but a second detector, near the neutrino production point is required. For this reason a near detector complex (ND280), at 280 meters from the beam production point, will be installed. In ND280 there will be several detectors inside the ex-UA1 magnet (see figure 1), that will provide a magnetic field of 0.2 T. The main purpose of ND280 is to measure the unoscillated neutrino spectrum and flux, the ν_e contamination and the interaction cross sections.

2. The T2K TPCs

In ND280 three large TPCs (see figure 2) together with 2 FGDs (Fine Grained Detectors), will provide a clean sample of charged current quasi elastic (CCQE) events. The momentum

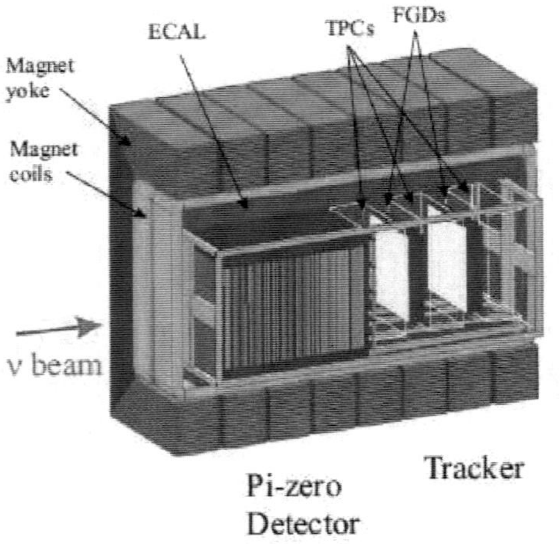

Figure 1. The ND280 complex

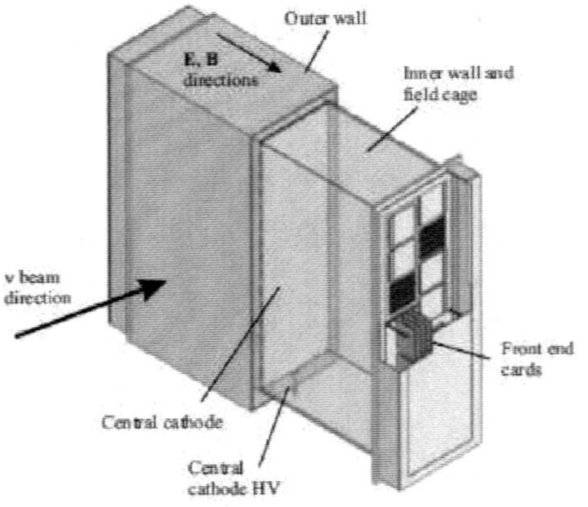

Figure 2. One TPC of the ND280 complex

of the muons produced in the interactions will be measured in the TPCs and will be used to determine the energy spectrum of the neutrino beam. Moreover in the TPCs it will be possible to distinguish electrons from muons measuring the energy released by the crossing particle in the gas. This will allow to measure the electron neutrino contamination in the beam.
The requirements for the T2K TPC can be summarized as follows:

- The momentum resolution must be better than 10% for 1 GeV muons. This requires a good point resolution due to the very low magnetic field.
- The energy scale has to be known at the 2% level. This can be achieved by an excellent control of magnetic and electric field distortions
- The dE/dx resolution has to be better than 10% to allow a 3σ separation of electrons and muons in the T2K energy region.

3. The TPC design

The ND280 tracker includes three identical rectangular TPCs consisting of an outer volume with a dimension of $2.5 \times 2.5\ m^2$ in the plane perpendicular to the neutrino beam direction, and 0.9 m along the beam direction. The active volume, equal to $1.8 \times 2.2 \times 0.7\ m^3$ is defined by the inner box divided into two halves by a central cathode to limit the maximum drift distance to 0.9 m. It will be filled with a gas mixture of $Ar/iC_4H_{10}/CF_4$ (95/2/3). A simplified layout of the TPC showing inner and outer boxes is presented in figure 2.
The central cathode is set at very high potential (close to 25 kV) and primary ionization electrons are drifted under an electric field of about 200 V/cm. The cathode will be loaded with aluminum strips for the laser calibration system. The outer box is separated from the inner box by a gap of 6.8 cm on the sides and top and 11.8 cm on the bottom. The surfaces of the outer box are aluminum at nominal ground potential. To guarantee the HV stability of up to 40 kV the gap between the inner and the outer box is filled with CO_2.
To amplify the signals and read out the TPCs, micro-pattern devices based on the Bulk-MicroMegas technology are used. This solution, developed by a CERN/TS-DEM and CEA Saclay Collaboration offers a good gas gain uniformity and allows to minimize the dead zones on the edges of the modules. The manufacturing process is described in detail in [4].

4. The Bulk MicroMegas

The Bulk-MICROMEGAS module designed for the T2K experiment consists of a segmented Printed Circuit Board used as anode covered by a photo-imageable polyimide film and a 430 LPI (lines per inch) woven micromesh, 30 μm thick. These three components are laminated together and undergo UV exposure with an appropriate mask. After a chemical development, the 20736 pillars with 400 μm diameter define the 128 μm amplification gap thickness. A 3.2 mm wide border at the edge made of a copper strip called BFM (border frame mesh) and a polyimide coverlay is also produced during the process (see figure 3). It allows to hold the stretched mesh without using an external additional frame. In this way, large surfaces can be instrumented with a compact, thin and robust low mass detector.

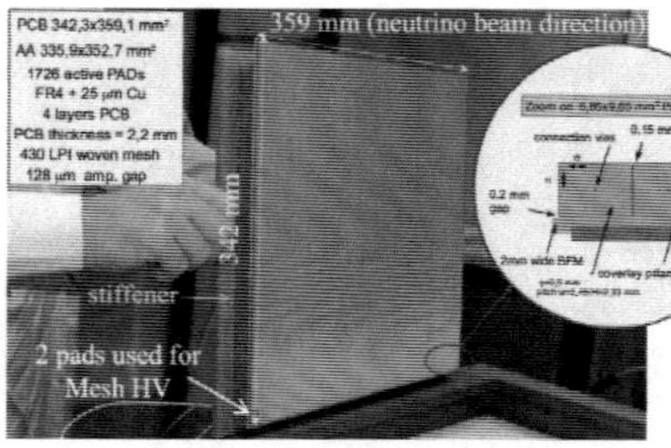

Figure 3. A 34 x 36 cm^2 Bulk-MicroMegas module glued on his stiffener

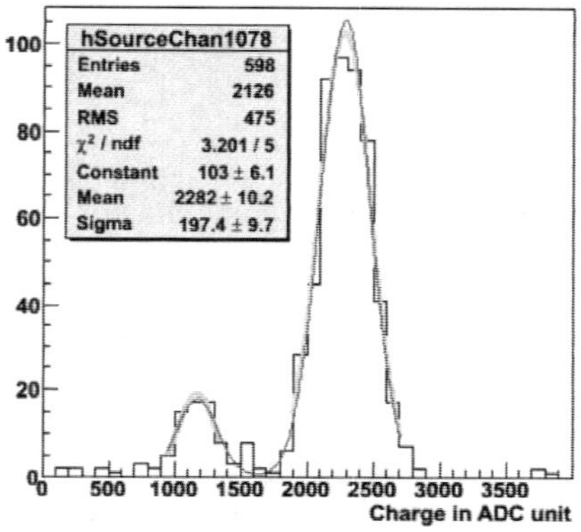

Figure 4. Charge spectrum in ADC unit reconstructed from the ^{55}Fe source in one pad

The detector is then glued on a FR5 mechanical frame with a ± 0.1 mm mechanical tolerance to stiffen the detection surface and to ensure good positioning, thereby ensuring a uniform drift electric field near the detection plane. The active area of a Bulk-MicroMegas module is 34×36 cm^2 and the anode is segmented into 1728 pads of 6.9×9.7 mm^2, arranged in 48 rows and 36 columns. Two pads located in one corner are used for the mesh high voltage connection from the backside of the PCB. A total of 72 modules are needed to instrument the three TPCs for a total effective surface of 9 m^2.

Each module is readout by six Front-End Cards (FECs) based on the low noise AFTER chip. This new ASIC has been designed specifically for TPC applications and its main characteristics are described in details in [5]. The AFTER ASIC consists of 72 multiplexed channels including for each channel a 12-bit amplifier/filter stage with 4 different charge ranges (120 fC to 600 fC) with various tunable peaking times (100 ns to 2 μs) and a Switch Capacitor Array (SCA) to sample and store the analog signal in 511 memory cells with a sampling frequency up to 100 MHz. The analog multiplexed output of the chip is read back and converted by an external ADC. A digital Front-End Mezzanine (FEM) card plugged directly on the six FECs is responsible for the slow control of the FECs, collection of digitized data, pedestals subtraction and zero suppression.

5. The test bench at CERN

Each Bulk-MicroMegas module is tested and validated with a dedicated test-bench located at CERN. The module is mounted on a gas box with an uniform electric field in a drift space of 4 cm. A 370 MBq ^{55}Fe source is placed on a motorized holder for a pad-per-pad scan over the whole active area.
The main purposes of the test bench are to find faulty pad, to provide absolute energy calibration and energy resolution measurement for each pad and to obtain the curve of the gain as a function of the applied voltage. These information are obtained by reconstruction of the 5.9 keV photoelectric peak spectrum of the ^{55}Fe source.
The scans are performed with the T2K/TPC front-end electronics and a drift electric field of about 200 V/cm is applied. The gas mixture used is $Ar/iC_4H_{10}/CF_4$ (95/2/3).
Up to now 12 modules have been tested in the test bench and only one faulty pad (over 21000) has been found. The energy resolution at 5.9 keV is of the order of 9% (see figure 4) with a dispersion of 6%. The pads have a good gain uniformity with a dispersion of 3% and the results obtained for the different modules show a good reproducibility in terms of gain uniformity and energy resolution.

6. Test with cosmic rays in the HARP TPC

In September 2007 one prototype of the MicroMegas detector for the T2K TPC was installed in the field cage of the HARP experiment [6]. During this test cosmic rays were taken with different magnetic fields (from 0 to 0.4 T) and different MicroMegas High Voltage and electronic settings.
The purpose of these tests was to study the performance of the MicroMegas prototype in terms of spatial and energy resolution. To study the energy resolution a truncated mean method is used.

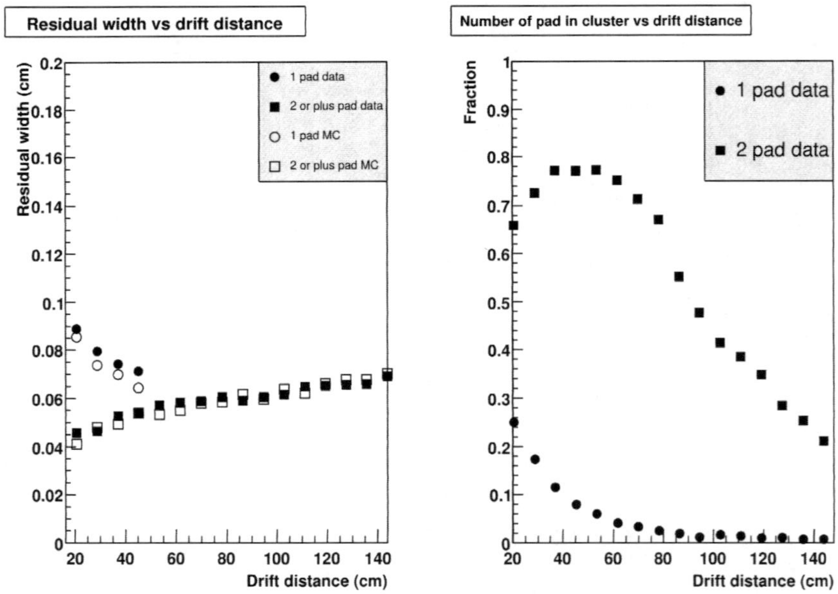

Figure 5. The measured spatial resolution (left) and number of pad fraction (right) for the test with cosmic rays in the HARP field cage (from [6])

With a gas mixture of $Ar/iC_4H_{10}/CF_4$ (95/2/3), with a magnetic field of 0.2 T and an electric drift field of 160 V/cm, a spatial resolution of 600 μm at 100 cm was measured (see figure 5), while the energy resolution, measured for cosmics with a momentum between 0.7 and 1 GeV, was $(12 \pm 1)\%$ for a track length of 36 cm.

7. Test Beam at TRIUMF

Between october and december 2008, beam tests with a mixed e^-, π and μ beam and the first T2K/TPC module have been performed. The TPC was equipped with 12 Bulk-MicroMegas modules and 4 of these modules were equipped with the front-end electronic. Data have been taken, without magnetic field, for various particle momenta, between 100 and 350 MeV/c. With a Time Of Flight system was possible to select samples of each particle, independently from the TPC response.
The MicroMegas modules and the electronic operated smoothly, without particular problems. The measured sparking rate was of 1 spark per hour per module at 360 V.

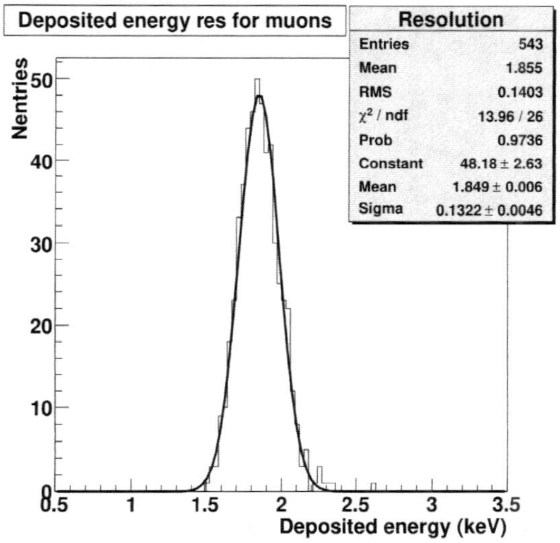

Figure 6. Deposited energy resolution for muons with a momentum of 150 MeV/c

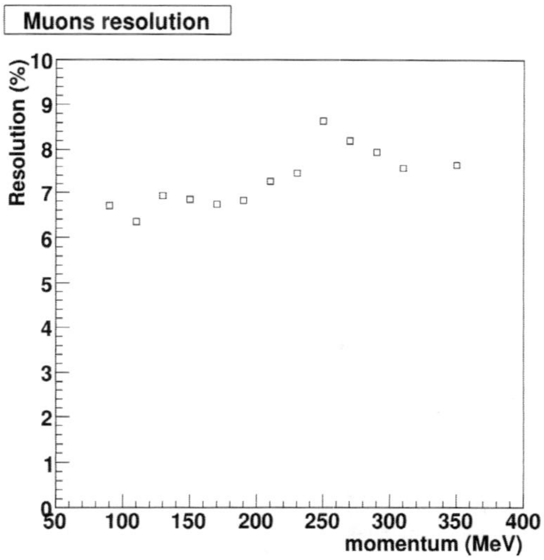

Figure 7. Deposited energy resolution as a function of particle momentum

One of the main goal of these tests was to study the particles identification capability of the TPC. In the TPC the particle identification is performed measuring the energy loss of the particle crossing the detector. To measure the energy loss a truncated mean method is used. For an horizontal track crossing the TPC, 72 deposited energy measurement will be provided, one for each MicroMegas column. The truncated mean method consists in the exclusion of a fixed fraction of clusters where the energy deposition is larger due to the Landau tails and, then, in the measurement of the mean energy deposited in the remaining clusters. The method was initially developed with a Monte Carlo simulation and it was found that, using the 70% of the columns with less charge,the resolution in the measurement of the energy loss was optimized.
Using the test beam data it was possible to measure the energy resolution of the TPC for the different samples of particles. In figure 6 it is shown the typical resolution for a sample of monochromatic muons, with a momentum of 150 MeV/c, that cross all the TPC. In this case the energy resolution was measured to be 6.9%. In figure 7 it is shown the energy resolution for muons as a function of the momentum of the particle. As it can be seen the energy resolution

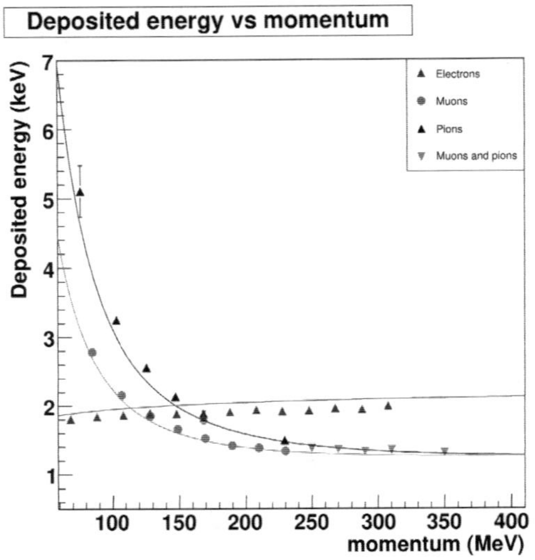

Figure 8. Measured deposited energy loss as a function of momentum for muons, electrons and pions, compared with a Monte Carlo simulation

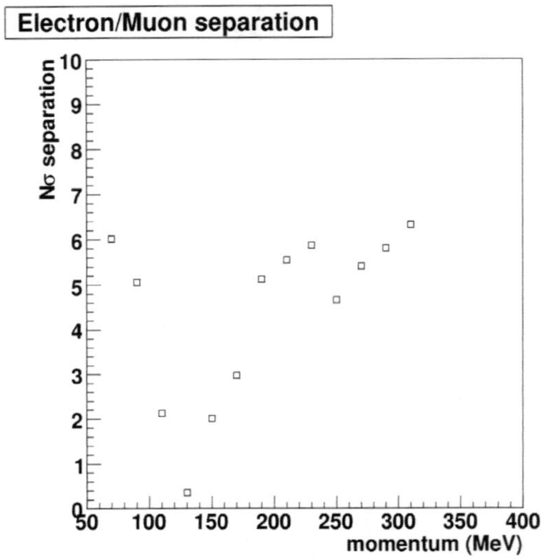

Figure 9. Measured separation between electrons and muons as a function of momentum

is better than 8% for each analyzed momentum.
The energy loss measured for the different particle samples, as a function of the momentum, is shown in figure 8 and it is compared with a Monte Carlo simulation of the energy loss curve. Finally also the electron/muon separation has been measured (see figure 9). A separation better than 5σ was observed if the particle momentum is larger than 200 MeV/c.

8. Conclusions

The T2K experiment will use three large TPCs based on large size Bulk-MicroMegas detectors as amplification devices. The production of the first TPC is completed and has been equipped with 12 Bulk-MicroMegas modules.
Calibration of the first 12 modules shows a good reliability of manufacturing process and good performance in terms of energy resolution and gain uniformity over the whole active surface.
Both, cosmic and beam tests, have demonstrated that the Bulk-MicroMegas technology is suitable to meet T2K requirements. In particular a spatial resolution of 100 μm at a drift distance of 100 cm and an energy resolution below the 8% for muons have been measured at momenta larger than 200 MeV/c.
The installation of the fully instrumented TPCs at JPARC will start in August 2009 while the T2K experiment will start the data taking in December 2009.

References

[1] Y. Hayato [T2K Collaboration], Nucl. Phys. Proc. Suppl. 143 (2005) 269.
[2] Y. Fukuda *et al.*, Nucl. Instrum. Meth. A 501 (2003) 418.
[3] M. Apollonio *et al.* [CHOOZ Collaboration] Phys.Lett. B466 (1999) 415-430
[4] I.Giomataris *et al.*, Micromegas in a Bulk, Nucl. Instr. Meth. A 560 405, 2006
[5] P. Baron *et al.*, AFTER an ASIC for the Readout of the large T2K Time Projection Chambers, IEEE NSS07 conference record N29-7 1865, 2007.
[6] M. Zito *et al.*, Large Bulk Micromegasdetectors for TPC applications, Nucl. Instr. Meth. Preprint submitted, 2008.

Journal of Physics: Conference Series **179** (2009) 012003 doi:10.1088/1742-6596/179/1/012003

Spherical TPC development and trends

S Andriamonje[1], S Aune[1], E Bougamont[1], M Chapellier[1], P Colas[1], J Derré[1], E Ferrer-Ribas[1], S Gaffet[4], G Gerbier[1], I Giomataris[1*], M Gros[1], I G Irastorza[5], P Mangier[1], R Lemrani[1], X F Navick[1], P Piquemal[3], P Salin[4], I Savvidis[2], J D Vergados[6] and M Zampaolo[3]

[1]CEA-Saclay
[2]University of Thessaloniki, Greece
[3]Laboratoire Souterrain de Modane
[4]Laboratoire Souterrain à Bas Bruit
[5]Laboratorio de Física Nuclear y Astropartículas, University of Zaragoza, Spain.
[6]University of Ioannina

Abstract. We present a new type of gaseous detectors, the spherical proportional counter. It consists of a small metallic ball, 16 mm in diameter, located at the centre of curvature of a large spherical vessel, acting as a proportional amplification structure. It allows high gas gains to be reached and operates from low to high gas pressure. This simple and robust structure allows large volumes to be read out with a single electronic channel. The detector performance presently achieved is already close to fulfil the demands of many challenging projects from low energy neutrino physics to dark matter detection with applications in neutron, alpha, gamma spectroscopy. Preliminary results of neutron flux measurements currently going on in underground laboratory will be presented.

1. Introduction

The Spherical Proportional Counter, recently developed, is a novel concept with very promising features, among which is the possibility of easily instrumenting large target masses with good energy resolution and low energy threshold. The natural radial focusing of the spherical geometry allows collecting and amplifying the deposited charges by a simple and robust detector using a single electronic channel to read out a large gaseous volume. Moreover the detector is robust and stable in time. Unlike in the conventional TPC special field cage is not needed.

There is increasing interest of such massive low-background, low-energy threshold detectors in particle astrophysics for searching the origin of Dark Matter in our universe or studying low-energy neutrino physics [1,4,5]. Large gaseous detectors located underground have been proposed to pursue such important new physics possibilities. In addition, for the search of Dark Matter in underground area, it is crucial to know with a good precision the contribution of the neutron background. At present a high efficiency and a good energy resolution neutron detector able to measure a very low neutron flux, expected to be at most 10^{-6} n/cm^2/s in a large energy range from thermal to several MeV, does

[*] Corresponding author. E-mail: ioa@cea.fr

not exist. The neutron detector based on the same geometry will use Helium-3 on a large mass scale. The resulting good resolution detector could provide the neutron energy spectrum at ground and underground installations. Good energy resolution is also required to obtain a low energy threshold and to identify lines from gamma background.

2. Detector description

A prototype of a spherical detector of 1.3 m diameter has been built, in the last two years at Saclay, as a demonstrator and its initial operation is quite promising. The design presents two basic innovations:

- A large spherical drift vessel, 1.3 meter in diameter and 6 mm thick
- A stainless-steel ball, 16 mm in diameter, is located at the centre of the drift vessel and is maintained by a metallic rod. Applying a positive voltage on the ball, while external sphere is grounded, a radial field is created.

The detector operates in a seal mode: the spherical vessel is first pumped out and then filled with an appropriate gas at a pressure from a few mbar up to 5 bar. In the simplest design the small spherical electrode acts as proportional amplification counter and is read by a single electronic chain. Figure 1 shows a schematic view of the device and describes the principle of operation; for more details we refer to previously published papers [2-3].

The amplification process is similar to other commonly used gaseous detectors: the Parallel Plate Avalanche Chamber (PPAC) [6] and the cylindrical proportional counter. Let us compare the three concepts in terms of electric field shape relevant for signal development and capacity relevant for electronic noise:

In the PPAC the electric field is constant and the capacity is given by $C = e_0 S / d$ where S is the detector surface and d is the thickness of the amplification gap. For large detectors (>1 m^2) the capacity, proportional to the surface, will exceed 10 nF.

In the cylindrical proportional counter the electric field is given by $E = V_0 / r$ where V_0 is the applied voltage and r is the radius of the sense wire. The capacity is given by $C = e_0 L \log(r / R)$ where L is the length of the sense wire and R the external radius; for large detectors we get $C >$ 100pF.

In the spherical proportional counter the electric field at a distance r is given by:

$$E(r) = \frac{V_0}{r^2} \frac{1}{1/r_2 - 1/r_1} \tag{1}$$

where r_2 is the anode radius, r_1 cathode radius and V_0 the anode voltage. The capacity is given by:

$$C = \frac{4\pi\varepsilon_0}{1/r_2 - 1/r_1} \tag{2}$$

and since ($r_1 >> r_2$) $C = 4\pi\varepsilon_0 r_2$.

For a typical size of r_2=6 mm the capacity is C = 0.05 pF. Such low capacity will provide a very low electronic noise and therefore low energy threshold even if the detector operates at moderate gains.

Journal of Physics: Conference Series **179** (2009) 012003 doi:10.1088/1742-6596/179/1/012003

Notice also that the capacity depends only on the size of the ball radius and is roughly independent of the vessel volume.

In a previous paper [7] we have presented results of high resolution energy measurements obtained by this counter. In the following section we will describe another application where the spherical detector is optimized for neutron measurements.

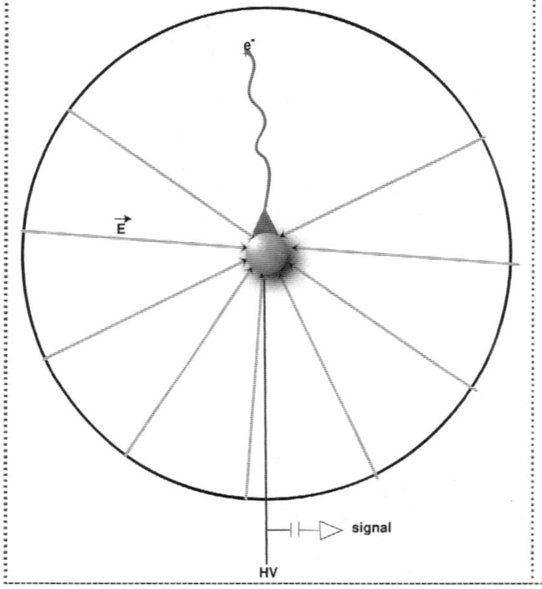

Figure 1. Principle of operation of the novel detector. Electrons produced by a charged particle in the gas vessel drift to the centre of curvature of the sphere under the action of the radial electric field. At a few mm of the small ball the electric field is intense and an avalanche is produced. Positive ions moving backward induce a signal to the preamplifier through a decoupling capacitor.

3. Neutron detection

Gas detectors are widely used as neutron detectors and are superior to scintillation counters in terms of stability, price per unit, efficiency and suitability for mass production. Gas-filled cylindrical detectors typically employ ^3He or BF_3 as the primary constituent, at pressures of less than 1 to about 20 atm. Our detector will operate with ^3He gas to capture and detect neutrons.

Neutrons interact with ^3He as follows:

$$n + {}^3He \Rightarrow p + {}^3H + 764 \text{ keV}$$

In order to measure the neutron energy, both the charged particles ^3H and p from the reaction must deposit all their energy in the drift volume. If the reaction takes place close to the vessel wall, it is likely for one of the charged particles or both of them to hit the wall and to loose a part of their energy. This is known as the wall effect and leads to wrong estimation of the neutron energy. The large volume with the possibility to operate at high gas pressure is an important advantage of our detector compared to the traditional cylindrical proportional counter.

Recently the detector has been installed in the LSM laboratory in Modane (Frejus). The goal was to measure thermal neutrons using 3 g of ^3He and evaluate the sensitivity for fast neutron measurements.

Figure 3 shows the amplitude of the signal versus the rise time (ms). The thermal neutron peak (ADC bin 1000) and the contamination due to ^{210}Po (ADC bin 4500) are clearly seen. To avoid such contamination the new detector will be fabricated by using low-radioactivity materials.

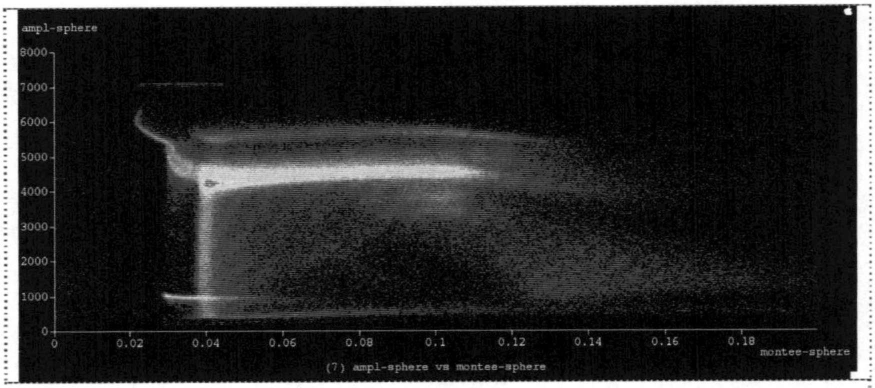

Figure 2.

Amplitude versus rise time: at bin 4500 is the amplitude of contamination due to ^{210}Po (5.3 MeV) while thermal neutron capture signal is at the bin 1000 (765 keV).

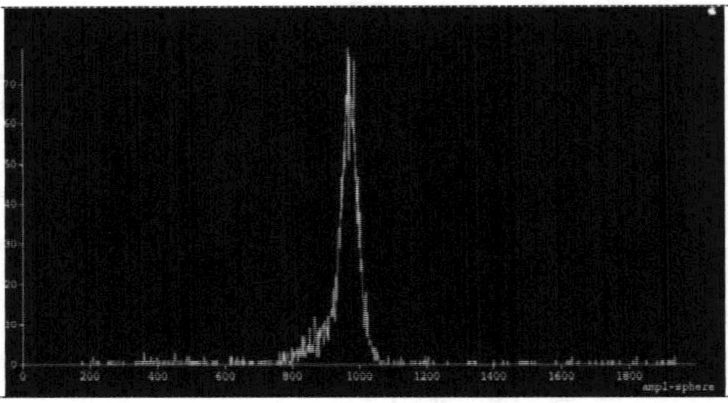

Figure 3

Pulse height of thermal neutron capture after rise time cut.

A pulse shape cut (rejecting long rise time pulses > 0.04 ms) keeps 60% of thermal neutrons but the effect on the background is spectacular (Figure 3); as expected background from gamma and beta particles is highly reduced and most of Po alpha rays emitted from the internal vessel surface (largest distance giving rise to long time pulses), are rejected.

The rate of selected thermal neutrons was 350/day. From this result we have a first estimate of the thermal neutron flux at LSM: 2 x 10^{-6}/cm^2/s.

Despite such powerful background rejection it will be difficult to measure fast neutrons with present apparatus; the cross section of fast neutron capture is three orders of magnitude lower. We need to build a new detector using low background materials and possibly an appropriate shield to reduce environmental backgrounds. Moreover we could anticipate by increasing the present (3 gr) amount of ^{3}He; by using 50 g of ^{3}He we expect a fast neutron capture rate of a few events a day.

4. Conclusions

We have developed a new detector based on the radial geometry with spherical proportional amplification read-out. The detector is robust, stable in time and is operating in seal mode. Another advantage of this structure is the use of a single electronic channel to read-out a large volume. This single information still allows the determination of the radial coordinate of the interaction point through the measurement of the time dispersion of the detected charge pulse. Such information is of paramount importance for localization of the interaction in depth and background rejection applying fiducial cuts.

Adding ^{3}He gas the present detector becomes a powerful massive neutron detector. Preliminary neutron measurements at underground with 3 gr ^{3}He gas are encouraging. Thermal neutron flux has been precisely measured and the detector operates in a stable fashion for many months. Prospects for measuring fast neutrons are open.

Several other applications are open arising from low energy neutrino physics, double beta decay to WIMP search.

5. References

[1] Giomataris I and Vergados J D 2004 *Nucl.Instrum.Meth.A* **530** 330-58
[2] Giomataris I et al 2006 *Nucl.Phys.Proc.Suppl.* **150** 208-13
[3] Aune S et al. 2005 *AIP Conf.Proc.* **785** 110-8
[4] Giomataris I and Vergados J D 2006 *AIP Conf.Proc.* **847** 140-6
[5] Giomataris I and Vergados J D 2006 *Phys.Lett.*B **634** 23-9
[6] Giomataris I 1998 *Nucl. Instr. and Meth. A* **419** 239.
[7] Giomataris I et al. 2008 *JINST* **3** P09007.

Journal of Physics: Conference Series **179** (2009) 012004 doi:10.1088/1742-6596/179/1/012004

Double beta decay in liquid xenon

Razvan Gornea on behalf of the EXO collaboration
LHEP, Bern University, Switzerland
E-mail: gornea@gmail.com

Abstract.
The Enriched Xenon Observatory (EXO) collaboration will search for double beta decay using the ^{136}Xe isotope [1]. During the initial phase (EXO-200), a 200 kg enriched Xe liquid Time Projection Chamber (TPC) will be deployed in the underground experimental area at the Waste Isolation Pilot Plant (WIPP), Carlsbad, USA. We present the design, construction and installation of the EXO-200 TPC focussing on the specific experimental requirements for the detection of neutrinoless double beta decay: high energy resolution and low radioactive background.

1. Introduction

The double beta decay, a rare nuclear process, is investigated experimentally because it has the potential to reveal new information about the nature and properties of neutrinos [2]. Indeed, this nuclear transition can proceed through multiple channels: with the emission of two electrons and two anti-neutrinos ($2\nu\beta\beta$) as predicted by the standard model of particles, without the release of neutrinos ($0\nu\beta\beta$) or accompanied by the radiation of a third light neutral fermion. The half-life of the neutrinoless decay depends on the effective neutrino mass, a linear combination of the three neutrino masses, and therefore its measurement provides further information about the neutrinos masses which cannot be independently established only from neutrino oscillation experiments [3, 4]. Furthermore, the observation of this exotic decay process would be a direct evidence that the neutrino is a Majorana particle. The spectrum of the energy carried by the two electrons is very different for the various decay channels, in particular for $0\nu\beta\beta$ this is simply a peak at the Q value of the decay. Therefore, events from each channel can be discriminated to a certain extent even when using detectors not sensitive to the topology of particle tracks.

The EXO R&D program aims at the deployment of a very large mass (ton scale), ultra low background TPC filled with xenon enriched in ^{136}Xe isotope for the detection of $0\nu\beta\beta$. For the first phase, the collaboration has successfully procured 200 kg of 80% enriched xenon that will be employed in a cryogenic liquid TPC providing charge and scintillation light readout. A high pressure gas phase TPC, to be implemented with larger xenon masses, is still under consideration for the next phases because it offers better spatial resolution facilitating efficient event selection based upon the topology of particle tracks. Also, future detection schemes will involve identification of the final state (i.e. $^{136}Ba^{+}$) which allows a drastic reduction of the effective radioactive background.

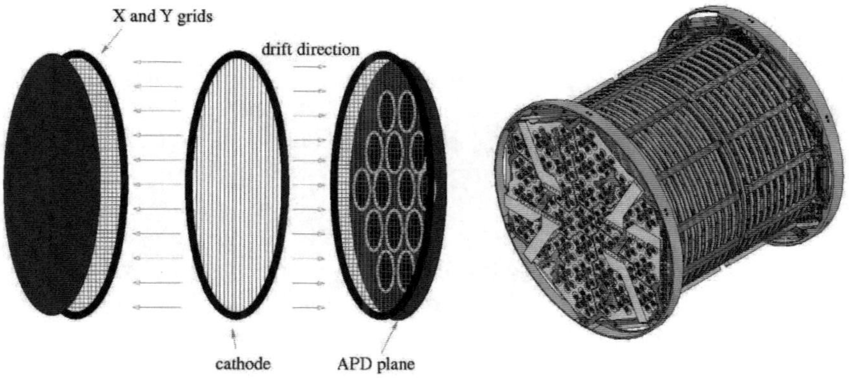

Figure 1. TPC diagram and 3-D drawing.

2. Experimental requirements

For the ^{136}Xe isotope, the energy released in the double beta decay ($Q = 2458$ keV [5]) is higher than that of most radioactive decays from the U and Th chains. Furthermore, xenon can be purified and enriched efficiently and has been used successfully as a detection medium with TPCs [6]. In the case of $0\nu\beta\beta$ decay, the radioactive background (i.e. events with energies around the Q value that emulate the expected signal) has to be kept under 40 events per year to optimize the EXO-200 detector performance and reach a half-life sensitivity of the order of 5×10^{25} years. Also, very strict radioactive background control is necessary to be able to observe the $2\nu\beta\beta$ decay that has a wide energy spectrum. It is therefore extremely important to reduce, or even eliminate, the residual radioactive impurities (from naturally occurring elements, especially K and those from the U and Th chains) found in all the components of the TPC and the surrounding experimental setup [7]. This can be achieved by careful selection of the construction materials and design optimization techniques intended to reduce the mass of the detector's components, especially those located close to its active volume [8]. Since, ultimately, the allowed double beta decay becomes the dominating background for the $0\nu\beta\beta$ mode, when considering a realistic detector with finite energy resolution, it is obvious that improving the energy resolution is an essential goal independently from the radioactive background control considerations.

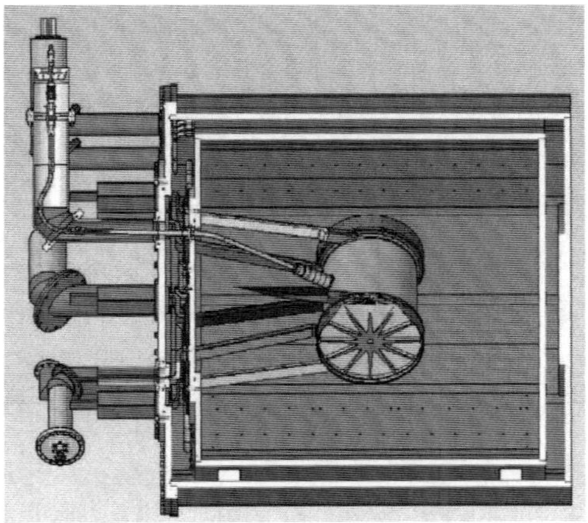

Figure 2. EXO-200 cryostat.

3. EXO-200 detector

The EXO-200 TPC has a cylindrical shape, with a diameter of 40 cm and a length of 35 cm, and it is segmented into two zones by a central photo-etched cathode made of phosphor bronze. Both regions are equipped with induction and charge collection wire grids followed by LAAPD (Large-Area Avalanche PhotoDiode) planes, placed parallel to the cathode at the ends of the chamber. Field shaping rings are supported by acrylic pillars and thin Teflon sheets, installed radially, serve as ultraviolet reflectors for the scintillation light. Figure 1 shows a diagram and a 3-D drawing of this assembly. The chamber is made of ultra low radioactivity copper formed from 1.5 mm thick rings welded together employing low contamination methods (e-beam and TIG welding) performed in controlled environments. Each LAAPD plane contains 250 UV ($QE > 1$ at 174 nm) sensors with an active diameter of 1.6 cm that are operated at about 1500 V for gains in the range 100× to 150×.

Figure 3. WIPP underground experimental area.

3.1. EXO-200 cryostat

The TPC is hosted in a refrigeration based cryostat that uses 4.2 tons of high purity heat transfer fluid. Due to safety constraints, liquid nitrogen cooling techniques cannot be employed at the WIPP underground facility and therefore a refrigeration based solution has been selected. Multiple heat transfer fluids have been evaluated and 3M's HFE-7000 was chosen primarily because it has the lowest residual radioactive contamination of all the candidates [8]. This fluid has a dual role, it serves as the inner gamma ray and neutron shield as well as the thermal bath that maintains the xenon temperature uniform. The cryostat has a cylindrical shape, with a diameter of 1.5 m and a length of 1.5 m, and it is made of ultra low activity copper. A 3-D drawing is shown in figure 2. This cryogenic scheme has been fully commissioned at Stanford University including the xenon and HFE handling systems.

3.2. Underground installation

The WIPP underground experimental area is located in a salt mine at a depth of 300 m. The EXO project has been allocated a vast dedicated tunnel that is provided with the necessary utility services. The collaboration installed a series of clean room modules supported by adjustable pillars in which the EXO-200 detector and its support infrastructure are enclosed, protected from mine dust and salt. Figure 3 shows a few pictures of the EXO experimental area before and after the placement of the clean room modules as well as during the installation of the cryostat and its outer lead shield. A muon veto composed of plastic scintillator panels (20 units with dimensions 65 × 315 cm^2 and 11 units with dimensions 65 × 375 cm^2) covers the module that hosts the detector. Monte Carlo simulations have been used to optimize the configuration of these panels and the best solution, illustrated in figure 4, provides a tagging efficiency of 99.7% resulting in the reduction of muon related background by a factor of 20×.

4. Radioactive background survey

The EXO collaboration conducted a large campaign to determine the residual radioactive contamination of many materials and components considered for the construction of the EXO-200 detector and its support infrastructure [8]. The potassium, thorium and uranium concentrations of more than 350 materials have been measured and a database containing the results for 225 interesting candidates has been made available to the experimental community developing detectors with similar low background requirements. Various methods have been employed for this survey: standard mass spectrometry (MS), glow discharge MS (GD-MS), inductively coupled plasma MS (ICP-MS), neutron activation analysis (NAA), alpha and gamma counting. To reach optimal sensitivity, each method imposes particular constraints on sample preparation, but generally those are complementary. For example, direct gamma counting offers the best sensitivity to cost ratio for large mass samples whereas ICP-MS performs the best for small mass samples that are chemically compatible with acid based preconcentration methods. A complete Monte Carlo simulation that includes the detailed geometry of the EXO-200 detector, the measured activities of the selected materials and event selection algorithms has been developed and allowed us to predict the background induced by residual contamination.

5. EXO-200 expected performance

The EXO-200 detector is expected to have a very low background induced by the residual radioactive contaminants, i.e. around 40 events per year in the energy range of interest for $0\nu\beta\beta$ decay. To reach such purity, it is necessary to do a careful selection of materials in parallel with a custom detector design. Also, all the manufacturing, handling and installation phases have to be done in clean environments to insure that additional radioactive contamination is not injected during these manipulations. An energy resolution of $\sigma = 1.6\%$ at 2.5 MeV is projected when combining both charge and scintillation light measurements. Therefore, considering reasonable parameters (200 kg of enriched xenon, 70% efficiency and 2 years run time), the expected sensitivity of EXO-200 is $T_{1/2}^{0\nu\beta\beta} = 6.4 \times 10^{25}$ which translates, in terms of effective neutrino mass to $\langle m_\nu \rangle = 133$ meV when the nuclear matrix elements are calculated with QRPA [9] (alternatively $\langle m_\nu \rangle = 186$ meV for calculations done with NSM [10]). Also, drawing advantages from the EXO-200 low radioactive background effort, the $2\nu\beta\beta$ decay mode of the ^{136}Xe isotope could be observed for the first time (current limit is at $T_{1/2}^{2\nu\beta\beta} > 1.2 \times 10^{24}$).

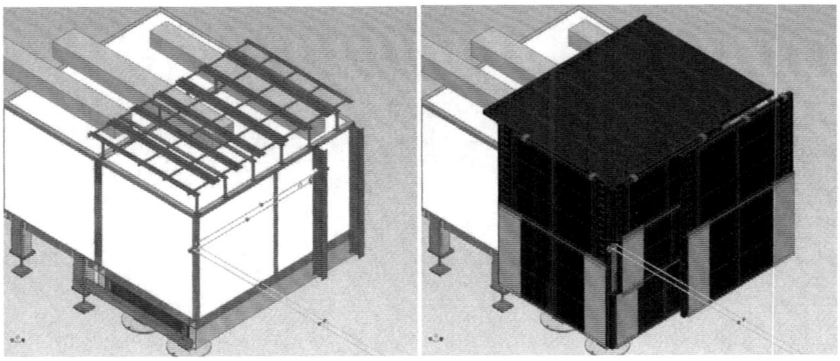

Figure 4. Muon veto panel formation.

6. Energy resolution study

A decisive requirement for $0\nu\beta\beta$ detectors is very good energy resolution because the ratio of signal to $2\nu\beta\beta$ induced background depends strongly on the aforementioned quantity:

$$\frac{S}{B} = \frac{m_e}{7Q\sigma^6} \frac{T_{1/2}^{2\nu\beta\beta}}{T_{1/2}^{0\nu\beta\beta}} , \quad (1)$$

where m_e is the mass of the electron, σ the energy resolution of the detector and $T_{1/2}$ the half-life of the respective decay modes. Therefore, the energy resolution achievable in liquid xenon has been studied using a small cell equipped with a UV sensitive PMT and a charge readout system [11]. The active volume had a cylindrical shape, with a diameter of 20 mm and a height of 6 mm, and a very thin ^{207}Bi source was attached to the cathode grid. The measurements clearly indicate that ionization and scintillation signals in liquid xenon are anti-correlated as illustrated in figure 5. The cell has been operated at multiple drift voltages and the results point out that this effect is due to the dependence of the scintillation yield on the amount of charge recombination. Therefore, simultaneous measurements of light and charge provide better energy resolution than any single approach as it is shown in figure 6 (σ_{min} is the minor elliptical axis of the 2-D peak as it can be seen in figure 5). From this study, when considering a small volume, the scaled energy resolution expected at the Q value for the ^{136}Xe decay is 1.4%. Careful considerations have been applied during EXO-200 design to insure optimal scintillation and charge collection schemes to provide similar energy resolution performance for a much larger detection volume.

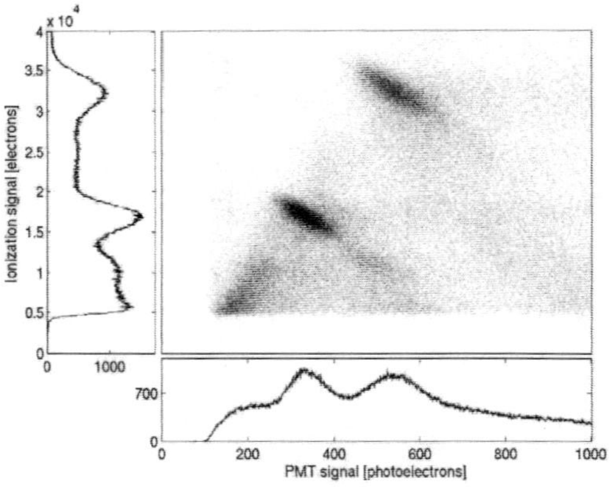

Figure 5. Two-dimensional scintillation and ionization spectrum (4 kV/cm drift field).

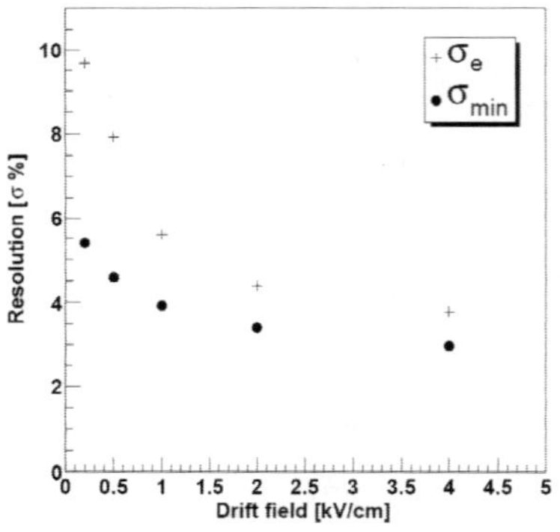

Figure 6. Energy resolution for the ionization channel alone compared to the minimal value obtained from a fit to the two-dimensional spectrum.

7. Detection of the final state

The double beta decay of xenon produces $^{136}Ba^{++}$ ions and their detection can be used as a very powerful tool to discriminate against radioactive background. Techniques using resonant light scattering from ions trapped in a RF cage have been applied successfully to Ba^{+} ions. Therefore, final state detection can be reduced to ion charge conversion and extraction from the detection medium followed by laser based spectroscopy. For the liquid phase, charge reduction occurs naturally because xenon has a larger ionization potential than barium and, for the gas

Journal of Physics: Conference Series **179** (2009) 012004 doi:10.1088/1742-6596/179/1/012004

phase, various additives can be employed with xenon. Ion extraction is a rather complex process especially for very large detectors and further R&D work is planned by the collaboration [12].

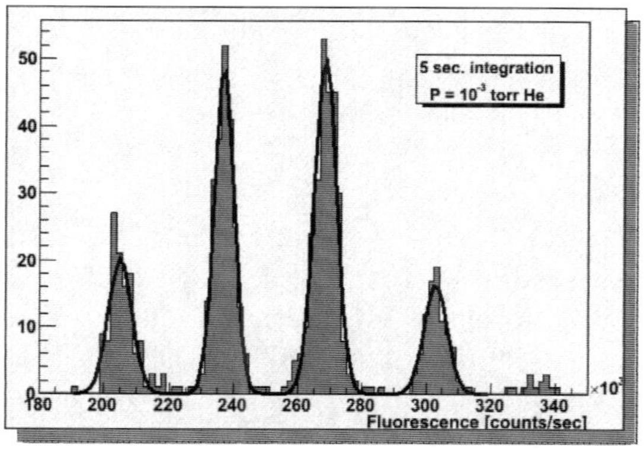

Figure 7. Histogram of the light scattered from barium ions.

Laser based Ba^+ tagging has been successfully accomplished in a RF cage using low pressure He ($P = 10^{-3}$ torr) cooling and reliable single ion transport has been demonstrated [13, 14]. Figure 7 shows the spectrum of fluorescence light for 5 s time slices, the first peak is associated with the background and the following peaks correspond respectively to 1, 2 and 3 ions (the area of each peak is proportional to the respective survival time). It illustrates the excellent resolving power of this method for counting ions, which is complemented by the capability of uniquely identifying Ba^+ ions (493.41 nm and 649.69 nm lasers are both required for resonant light scattering, the readout is performed in the blue frequency range and the red laser is used as a switch). In-situ solutions for liquid and gas phases are explored by the collaboration because they offer the advantage of dispensing with the extraction process.

8. Conclusion

The EXO-200 detector will be commissioned for underground operation during 2009. The collaboration has successfully conducted a R&D program to produce 200 kg of enriched xenon and designed a liquid phase TPC optimized for double beta decay search. Very low radioactive background and high energy resolution are expected. In the future, the detection scheme may be extended to include final state tagging to improve background rejection.

References

[1] Akimov D et al. 2005 *Nucl. Phys.* B **138** 224-6
[2] Avignone F T et al. 2008 *Rev. Mod. Phys.* **80** 481-516
[3] Giunti C 2000 *Nucl. Phys.* B **87** 1-3 324-6
[4] Amsler C et al. 2002 *Phys. Lett.* B **545** 57-61
[5] Redshaw M et al. 2007 *Phys. Rev. Lett.* **98** 053003
[6] Luescher R et al. 1998 *Phys. Lett.* B **434** 407
[7] Danilov M et al. 2000 *Phys. Lett.* B **480** 1-2 12-18
[8] Leonard D S et al. 2008 *Nucl. Instrum. Meth.* A **591** 490-509
[9] Rodin V A et al. 2007 *Nucl. Phys.* A **793** 1-4 213-5
[10] Caurier E et al. 2008 *Phys. Rev. Lett.* **100** 052503
[11] Conti E et al. 2003 *Phys. Rev.* B **68** 054201
[12] Fierlinger P et al. 2008 *Rev. Sci. Instrum.* **79** 045101
[13] Green M et al. 2007 *Phys. Rev.* A **76** 023404
[14] Flatt B et al. 2007 *Nucl. Instrum. Meth.* A **578** 399-408

The NEXT experiment

J. Díaz[1], N. Yahlali[1], M. Ball[1], J. A. S. Barata[7], F. I. G. M. Borges[7], E. Calvo[2], S. Cárcel[1],J. M. Carmona[4], S. Cebrián[4], A. Cervera[1], X. Cid[9], C. A. N. Conde[7], T. Dafni[4], T. H. V. T. Dias[7], L. M. P. Fernandes[7], E. Ferrer-Ribas[8], E. D. C. Freitas[7], J. Galán[4],A. Gil[1], I. Gil[5], I. Giomataris[8], H. Gómez[4], J.J. Gómez-Cadenas[1], F. Granena[2], J. A. Hernando-Morata[9], F. J. Iguaz[4], I. Irastorza[4], J. A. M. Lopes[7], D. Martínez[9], C. M. B. Monteiro[7], J. Muñoz-Vidal[1], C. Palomares[5], I. Irastorza[4], M. Lázaro[1], T. Lux[2], G. Luzón[4], J. Martín-Albo[1], F. Monrabal[1], J. Morales[4], F. Nova[2], P. Novella[2], D. Nygren[3], L. Ripoll[6], A. Rodríguez[4], J. Ruz[4], N. F. Sánchez[2],F. P. Santos[7], J. M. F. Dos Santos[7], L. Serra[1], M. Sorel[1], L. M. N. Távora[7], A. Tomás[4], J. F. C. A. Veloso[7], A. Verdugo[5].

[1]Instituto de Física Corpuscular (Centro mixto UV-CSIC), Apdo. de Correos 22085, E-46071 Valencia (Spain)
[2]Instituto de Física de Altas Energías, IFAE, Barcelona, Spain
[3]Lawrence Berkeley Laboratory, Berkeley, EEUU
[4]U. Zaragoza,Spain
[5]CIEMAT ,Spain
[6]U. Politécnica de Valencia, Spain
[7]Universidade de Coimbra, Portugal
[8]CEA, IRFU, Saclay, France.
[9]Universidade de Santiago de Compostela, Spain.

E-mail: `jose.diaz@uv.es`

Abstract. Neutrinoless double beta decay measurements are the most promising experiments both to reveal the Majorana nature of the neutrino and to set a value for its mass. The NEXT project propose to build a High pressure Xenon TPC in the Canfranc Underground Laboratory (Huesca, Spain) to measure double-beta decay of ^{136}Xe, both normal and neutrinoless, with a source mass of 100 kg of enriched xenon.

1. Introduction

The neutrino nature is one of the problems of Physics still unsolved. There exist experimental evidence that neutrinos oscillate [1], which imply they are massive. Mass differences between the different neutrino flavours are extracted from neutrino oscillation data. However, the mass of the lightest neutrino is still unknown, and only an upper limit of about 500 meV is established by different experiments and cosmological considerations. Double beta decay is a rare decay process which can be observed only if the corresponding single beta decays are energetically forbidden [2]. Only a few isotopes fulfill this condition. Normal double beta decay ($\beta\beta 2\nu$) occurs as a two-step process of ordinary beta decay, and produces a continuous energy spectrum of the electron pair, as the one shown in Fig. 1. However, massive neutrinos could be Majorana particles, which

Journal of Physics: Conference Series **179** (2009) 012005 doi:10.1088/1742-6596/179/1/012005

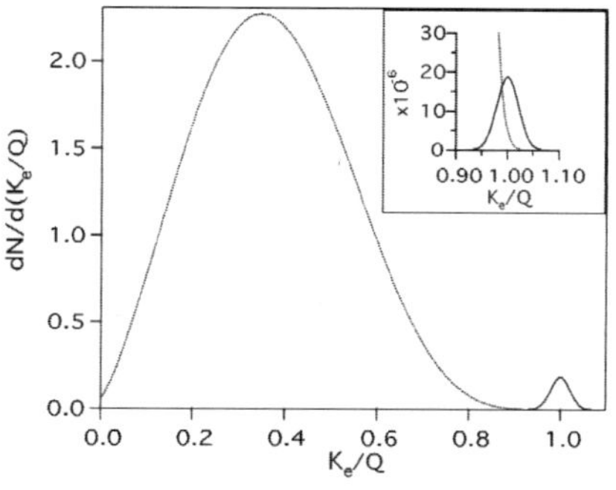

Figure 1. Double beta decay spectrum. The continuous part is the spectrum of the $\beta\beta2\nu$ process and the peak to the right correspond to the $\beta\beta0\nu$ process. The inset illustrates how the $\beta\beta2\nu$ process contribute as a background for the $\beta\beta0\nu$ peak for finite energy resolution. Taken from Ref.[3]

implies they are identical to their corresponding antineutrinos. In this case, double beta decay could take place without emission of neutrinos and the ideal energy spectrum would be a spike at the $Q_{\beta\beta}$ value. The half-lives of double beta decay can be written for the neutrinoless and two neutrino cases as,

$$T_{1/2}^{0\nu\ -1} = G_{0\nu}(Q_{\beta\beta}, Z)\, [M_{0\nu}]^2\, m_{\nu\nu}^2 \tag{1}$$

$$T_{1/2}^{2\nu\ -1} = G_{2\nu}(Q_{\beta\beta}, Z)\, [M_{2\nu}]^2 \tag{2}$$

where $G_{0\nu} and G_{2\nu}$ are exactly calculable phase space factors, $M_{0\nu} and M_{2\nu}$ nuclear matrix elements, at present strongly model dependent, and $m_{\nu\nu}$ the electronic neutrino mass. The measurement of neutrinoless double-beta decay ($\beta\beta0\nu$) by any of the current or future experiments could show the Majorana nature of the neutrino and provide the value of its mass at the same time. However, to observe the $\beta\beta0\nu$ peak, one has to avoid any background in the $Q_{\beta\beta}$ energy region, including the $\beta\beta2\nu$ spectrum. In practice, this requires not only to work in a deep underground laboratory, and with very low levels of radioactivity in all the components the detector is made of, but also to measure electron energies with a resolution good enough to make the $\beta\beta2\nu$ background negligible in the region of the $\beta\beta0\nu$ peak.

Different double-beta decay experiments have been carried out during the past four decades [4], without any observation of $\beta\beta0\nu$ events up to the date (with the exception of the controversial data of the Heidelberg-Moscow collaboration known as Klapdor claim [5]), but have established more stringent limits to the neutrino mass. Double-beta decay detectors can be classified into two different types: calorimeters and track-calorimeters. The experiments of the calorimeter type measure the total kinetic energy of both electrons with state of the art resolutions. To these kind of experiments belong those based on Ge and bolometers. They have excellent energy resolution, which in any case is an essential characteristic of this method in order to separate true events from the different sources of background. As the source isotope belongs to the detector material, these detectors have better efficiency than detectors with an external source, due to the absence of windows and backscattering.

The track-calorimeter detectors measure both the energy of the electrons and some tracking information of the β particles. This tracking information provides a topological signal characterizing two electron events, allowing to discriminate these events from other background events like e-γ events. This is the case of the NEMO3 detector[6] which combines exclusive measurement of electron energies with tracking information sufficiently good to measure angular distributions[7]. In this detector, source isotope and calorimeter material are different. To avoid critical loss of efficiency through backscattering of the electrons in the calorimeter, NEMO3 (and its possible extension SuperNEMO) is limited to calorimeters of very low Z, which in practice means the use of plastic scintillators, characterized by a poor energy resolution (8% at best).

As the topological signal given by tracking information can produce a background suppression of about two orders of magnitude, the ideal detector should combine an energy resolution good enough to avoid $\beta\beta2\nu$ background under the $\beta\beta0\nu$ peak and a reasonable tracking information, allowing to reject background produced by non $\beta\beta$ events. Another important issue is scalability. If the $\beta\beta0\nu$ decay is not found in current experiments, detectors of larger masses (multiton) may need to be built. To obtain full advantage ofan eventual increase of mass, it is essential that background be proportionally reduced [4]. This is only possible when the source isotope is a volume which can be produced and kept along its running life in extraordinary conditions of radio-purity. A continuous purification system can be implemented only for liquids and gases.

2. The Canfranc Underground laboratory

The Canfranc Underground Laboratory (LSC) is a new underground facility located in the Spanish part of the Somport road tunnel connecting Spain and France. It is managed by a consortium of the University of Zaragoza, the Gobierno de Aragón (local government of the Aragón Spanish autonomous community) and the Spanish Ministry of Science and Innovation (MICINN). Its depth and cosmic muon flux relative to other underground laboratories is given in Fig. 2. In Fig. 3 the experimental hall A is shown.

The physics at LSC has been supported by the MICINN in the framework of the program CONSOLIDER-INGENIO, with the approval of the project CUP (Canfranc Underground Physics) in the 2008 opening of that program.

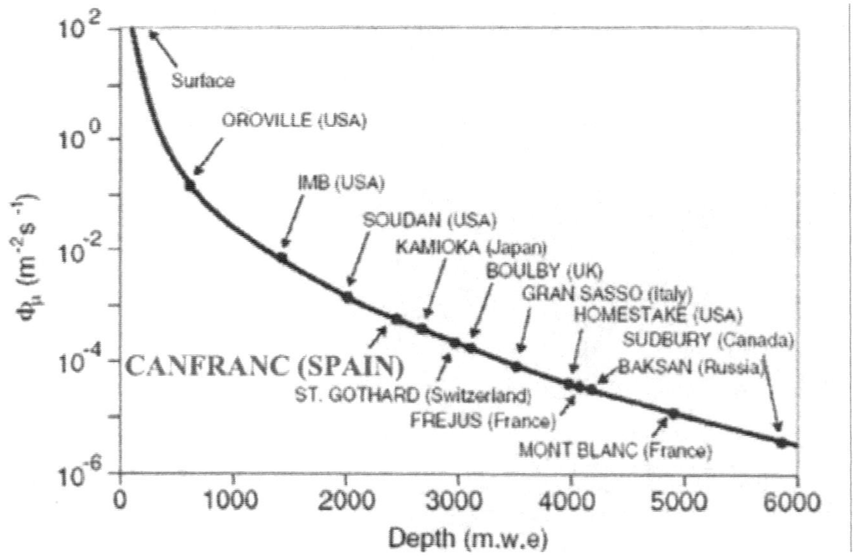

Figure 2. Dependence of the muon flux with the depth of different underground facilities, including the LSC.

3. The NEXT Experiment

The NEXT (Neutrino Experiment with a Xenon TPC) experiment is one of the proposals of CUP (in fact the main one) and has as a goal the design and construction of a High Pressure Xenon

Figure 3. Hall A of the LSC

TPC enriched in ^{136}Xe to measure its double beta decay both with neutrinos and neutrinoless. None of these modes have been observed yet. The NEXT collaboration is integrated at present by the Spanish universities of Girona, Santiago de Compostela, Politécnica de Valencia, Valencia and Zaragoza, and the research institutions CIEMAT (Madrid), Coimbra (Portugal), CSIC, IFAE (Barcelona), Saclay (France) and LBL (USA). Other institutions may join in the future.

The election of ^{136}Xe as source isotope has several obvious advantages: 1) ^{136}Xe has a large abundance in natural Xe (8.9%) and can be easily enriched by centrifuging; 2) Xe gas has no radioactive natural isotope and can be continuously purified by standard gas systems and kept to negligible levels of radioactivity; 3) xenon responds with a scintillation light ($\lambda = 175$ nm) to the passage of fast charged particles, which can be used as a trigger signal to measure double-beta decay events; 4) a TPC can work with pure xenon avoiding all the complications associated with gas mixtures; the failure to observe $\beta\beta2\nu$ decay by the Gotthard experiment [8] indicates that this process is suppressed by an unknown mechanism in ^{136}Xe and so produces a small background in the region of the $\beta\beta0\nu$ peak; and 5) $Q_{\beta\beta} = 2.4578$ MeV [9] for ^{136}Xe, which is a value reasonably high to have a reduced background.

The election of high pressure Xe TPC has the advantage over of liquid Xe TPCs, as those of the XENON [10] and EXO [11] experiments, the possibility of recording tracking information. The electron energies are deposited in a small region in the case of liquid Xe. In the case of gaseous Xe, the energy is deposited in a track of length dependent on the gas pressure. Simulations indicate that at a pressure of about 10 bars, a good compromise is obtained between efficiency and sufficient tracking length to provide a topological signature allowing to reject effectively background. At higher pressures, energy deposition tracks are too short to be used for background rejection. Also, the energy resolution may be better in gas than in liquid Xe.

The NEXT collaboration aims to reach mass of enriched Xe of 100 kg in a time schedule of about 5 years from this writing. Previous to the building of this detector, called NEXT-100, a smaller detector of about 10 kg of mass, called NEXT-10, will be build to check all the relevant issues: radiopurity, readout system, gas purification, electronics and data acquisition. NEXT-10 I is foreseen to be built in a time schedule of 3 years.

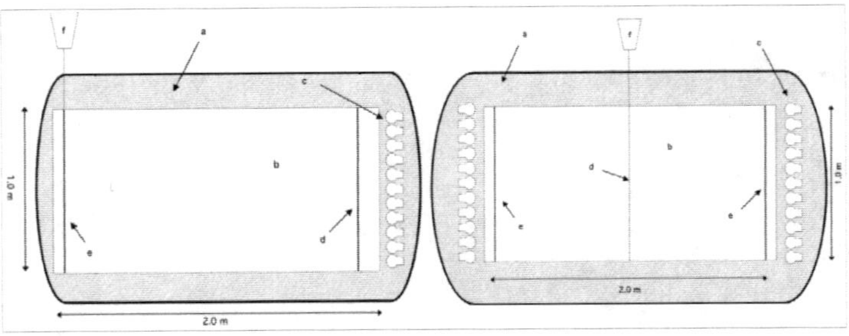

Figure 4. Concept of NEXT: A TPC of about 2 m length and 1 m diameter. The lamps indicate a readout system.

In Fig. 4 the concept of NEXT-100 is shown. The TPC will be a cylinder of about 2 m length and 1 m diameter. The cathode can be placed in one side or in the middle. This second option would allow to read on both sides. This could be necessary if different readout methods are used for energy and tracking readouts. For example, it could be possible to read the energy signal with photomultiplier tubes (PMT) on one side and the tracking pattern with silicon photomultipliers or Micromegas detectors on the other side.

The topological signal produced by two electron tracks is a kind of rope with two balls at its ends, as shown in the left part of Fig. 5. This track reminds the boleadoras employed to hunt by ancient natives of South America like the one depicted in the right part of the picture. The blobs are produced by the Bragg peak of the electrons as they are stopped in the gas. The track is not a rectilinear path but a zigzagging track due to Coulomb scattering. For this reason, no information about angular distributions can be obtained from tracking information. However, this topological signal is essential to distinguish two electron tracks from other kind of events, like the one electron track shown on the left part of Fig. 5. The tracking system finally selected should have a granularity high enough to distinguish the Bragg blobs produced by electrons stopping in the gas.

3.1. Readout technologies

At present, three different readout technologies are being considered. On one side, charge amplification readout by Micromegas is being developed by the Zaragoza and Saclay groups. Excellent energy resolution has been obtained with different gas mixtures, but further investigations are needed to be able to operate Micromegas in pure xenon at high pressure. On the other side, readout of scintillation light produced by electroluminiscence is being studied both with avalanche photodetectors (APD) and photomultiplier tubes. Both PMT and APD have been shown to deliver signals with excellent energy resolution (better than 1%) [11, 12]. Moreover, both APD and PMT can be produced with very good radiopurity levels. In the R&D phase we are going to use the PMT R8520 from Hamamatsu Photonics, already selected and tested by the XENON collaboration. These PMT do not stand pressures higher than 5 bar. In some prototypes, the PMT will be placed outside the TPC, and will read the scintillation light through quartz windows with high transparency in the $\lambda = 175$ nm region, and thick enough to stand 10 bar of pressure.

Different prototypes are going to be built to investigate the performance of each technology for high pressure pure xenon. The final decision will be taken in function of the tracking and energy resolution obtained, without forgetting the necessity of producing the different components with the required level of radiopurity. In Fig. 6, a prototype of a TPC read out by 16 PMT placed in side box is shown. The prototypes will work coupled to a vacuum system able to reach 10^{-7} torr, to clean all the impurities that could produce attachment in the gas. Once the vacuum is reached,

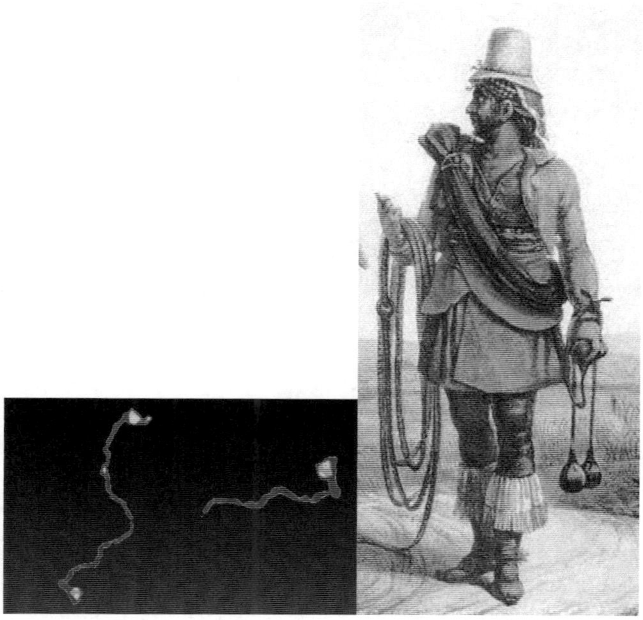

Figure 5. Topological characteristics of 2 electrons and 1 electron tracks. Two electron tracks are characterized by a boleadoras pattern (a rope with two balls like those carried by the indian native to the right)

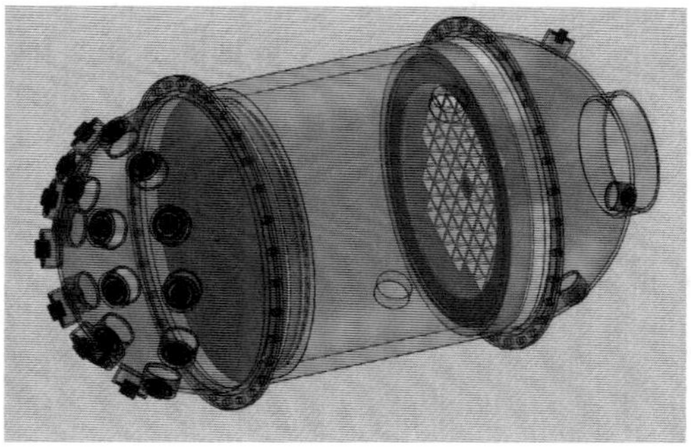

Figure 6. Prototype to test the read energy with PMT and tracking with SiPMTs

the gas system sketched in Fig. 7 enters into work.

3.2. Electronics

Although no decision has be taken yet concerning electronics and data acquisition, going fully digital seems the best option. Continuous sampling by flash ADC avoids the use of triggers, which always have an efficiency less than unity. As an additional advantage, a more stable electronics running is obtained. The big amount of data recorded can be reduced to reasonable amount by modern techniques of data compressing like those employed by the ICE CUBE collaboration. All the sampled values less than a given threshold are set to 0, and a zero suppression technique is applied. For the small prototypes, and probably for NEXT-10, commercial VME flash ADC available from CAEN will be used. For NEXT-100, the adaptation of one of the available chips is under study.

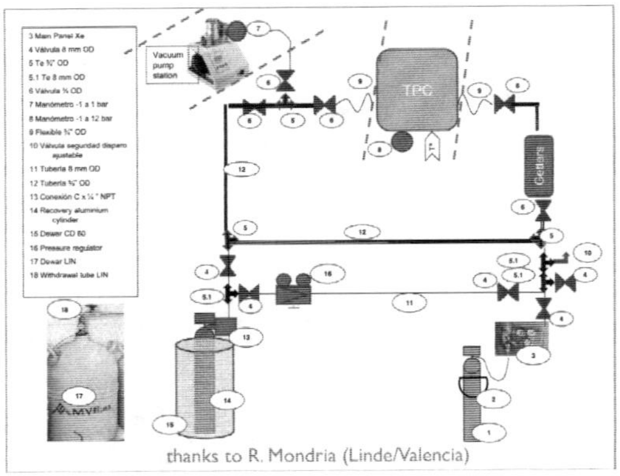

Figure 7. Sketch of the gas system at IFIC-Valencia

3.3. Background, Shielding and Radiopurity program

The most concerning background events for the NEXT experiment are electrons and gamma rays (which produce Compton electrons) with an energy above the Q_{energy}, 2.48 MeV. Most of these particles are originated from the ^{238}U and ^{232}Th chains coming from detector components and from the laboratory natural radioactivity. ^{208}Tl produced in the ^{232}Th chain, emits a photon of 2614.5 keV in the 99% of the decays, never alone, but together with several photons and electrons. To enter into the ROI this photon has to lose a fraction of its energy by bremsstrahlung radiation of the electrons produced in photoelectric absorption, which escapes the detector, or successive Compton scatterings in which low energy gamma rays escape from the detector. In the ^{238}U chain, two beta decays with Q_{max} above 2.48 MeV are produced: 2662.68 keV (1.7% of intensity) and 3272 keV (18.2% of intensity) from ^{214}Bi, ionizing always a few millimeters near the vessel surface, and a photon of 2447.8 keV (with an intensity of 1.57%) which follows a beta decay with Q_{max}=824.3 keV. Another concerning contamination is radon gas. It can be translated into ^{214}Bi, for the ^{222}Rn, or into ^{208}Tl, for the ^{220}Rn, their only decay products affecting the experiment through deposition on the surface of the readout system. Thanks to the use of la 15 cm lead shielding and muon veto, we can get external background reductions of the order of 10^{-5}. An active veto will also eliminate any event depositing its energy in the first centimeter of the vessel surface. Moreover, NEXT has excellent capabilities to get further reductions due to its good energy resolution (a FWHM=1% will eliminate the disturbing double beta contamination and part of the gamma contribution) and, very important, thanks to a good spatial resolution. The possibility to reconstruct electron tracks will eliminate multiple events, as those caused by multiple Compton interactions, and some single electron event (a reduction of the order of 10^{-2} according to Gotthard experiment). Monte Carlo simulations show that after a complete off-line analysis, we could easily, at this preliminary study, reach contamination levels below 0.01 counts keV^{-1} kg^{-1} year^{-1} (Gottard final level, with FWHM of 5.2% at 2480 keV, to be compared to the expected 1% for NEXT) which implies around 25 counts per year for 100 kg of ^{136}Xe, still a large number. The selection of high radio-pure materials, and an improvement of pattern recognition may quickly decrease this number.

The main contaminants that could fake the $\beta\beta$ signal are γ rays of energy higher than the $Q_{\beta\beta}$ value for ^{136}Xe. Gamma rays of these energies can be produced by one of the few cosmic rays reaching the Canfranc experimental hall by different mechanisms, in particular by spallation of rock materials and subsequent capture of the neutrons produced. This mechanism is particularly dangerous if neutrons reach the inner part of the detector and are captured there, because, in this case, pair converted γ rays in the energy window of $\beta\beta$ events cannot be distinguished of true events. External γ rays may also convert in the detector, but they can be strongly suppressed with moderate lead or iron shielding. Neutron shielding is more troublesome, requiring a big volume of hydrogen rich material like water. Calculation of the shielding needed, taking into account the LSC muon flux of $5{\cdot}10^{-3}$ μ cm^{-2}s^{-1} are being carried out by the U. of Zaragoza group.

4. Summary

The NEXT project is already approved by the Scientific Committee of the LSC and partially funded. In three years from now, a 10 kg detector working with natural xenon, NEXT-10, will be built. This detector will allow us to check all the technological issues and radioactive backgrounds. In 5 years a 100 kg detector working with xenon enriched in ^{136}Xe is intended be built. This NEXT-100 detector offers a new physics potential, since none of the double beta decay modes of ^{136}Xe has been measured up to now. The technology seems scalable to much larger masses, and in case the $\beta\beta0\nu$ process has not been discovered by then, a detector housing a mass of 1 ton or more enriched xenon could be envisaged.

Acknowledgments

This work has been supported in part by the Spanish Ministry of Science and Innovation under contract FPA2006-12120-C03. Part of this funding comes from FEDER funds.

[1] Kajita T 2006 *Rep. Prog. Phys.* **69** 1607
[2] McKeown RD and Vogel P 2004 *Physics Reports* **394** 315
[3] Elliot SR and Vogel P 2002 *Annu. Rev. Nucl. Part. Sci.* **52** 115
[4] Avignone FT, King GS and Zdesenko Yu G 2005 *New Journal of Physics* **7** 6
[5] Klapdor-Kleingrothaus HV *et al* 2001 *Eur. Phys. J.* A **12** 147
[6] Arnold R *et al* 2005 *Nucl. Instr. and Meth.* A **536** 79
[7] Arnold R *et al* 2005 *Phys. Rev. Lett.* **95** 182302
[8] Vuilleumier J.-C. *et al* 1993 *Phys. Rev.* D **48** 1009
[9] Redshaw M, Wingfield E, McDaniel J, and Myers EG 2007 *Phys. Rev. Lett.***98** 053003
[10] Aprile E *et al* 2005 *New Astronomy Reviews* **49** 289
[11] Akimov D *et al* 2005 *Nucl. Phys.* B (Proc. Suppl.) **138** 224
[12] Coelho LCC, Lopes JAM, Covita DS, Conceição AS, dos Santos JMF 2007 Nucl. Instr. and Meth. **575** 444

Journal of Physics: Conference Series **179** (2009) 012006 doi:10.1088/1742-6596/179/1/012006

Topological detection of double beta decay with NEMO3 and SuperNEMO

Ruben Saakyan, on behalf of the NEMO3 and SuperNEMO collaborations.

Department of Physics and Astronomy, University College London, Gower Street, London WC1E 6BT, UK.

E-mail: `saakyan@hep.ucl.ac.uk`

Abstract.
Since 2003 NEMO3 experiment has been searching for neutrinoless ($0\nu\beta\beta$) and two neutrino ($2\nu\beta\beta$) double beta decay using about 10 kg of enriched isotopes. A limit on the Majorana effective mass ($< m_\nu >$) has been obtained and several measurements of $2\nu\beta\beta$ decays have been performed. The results are reported here. In parallel, there is an active R&D programme for the SuperNEMO experiment which is expected to commence data taking in 2012-2013 and will use 100 kg of enriched isotope to reach a sensitivity to $< m_\nu >$ at the level of 50−100 meV by 2017. The unique approach of these experiments is a tracking plus calorimetry technique which allows a topological signature of $\beta\beta$ events to be detected.

1. Introduction

Neutrinoless double beta decay ($0\nu\beta\beta$) leads to the decay of a nucleus of charge Z and atomic number A via the process $(Z, A) \rightarrow (Z+2, A) + 2e^-$. This decay violates total lepton number L and its observation would therefore be a direct indication for physics beyond the Standard Model. Neutrinos would be Majorana and not Dirac particles, thereby solving one of the fundamental questions of particle physics. The $0\nu\beta\beta$ half-life is given by

$$[T^{0\nu}_{1/2}(A, Z)]^{-1} = |\eta|^2 \cdot |M^{0\nu}(A, Z)|^2 \cdot G^{0\nu}(Q_{\beta\beta}, Z), \quad (1)$$

where η is a lepton number violating parameter, $M^{0\nu}$ is the nuclear matrix element (NME) and $G^{0\nu}$ is a known phase space factor. The mechanism most commonly discussed is the one shown in Fig. 1 where a light Majorana neutrino is exchanged in which case $|\eta| = \langle m_\nu \rangle$. However many other mechanisms are possible: V+A currents, Majoron emission, R-parity violating SUSY, to name a few.

Observation of the topological signature of the $\beta\beta$ decay, the angular distribution between the emitted electrons and their individual energies provides a way of discriminating the underlying physics mechanism. In addition the $\beta\beta$ topological signature observed in a tracking detector provides an unambiguous signature of the process ("smoking gun"). The unique feature of the NEMO3 and SuperNEMO experiments is a tracking plus calorimetry technique which allows a topological signature of $\beta\beta$ events to be detected.

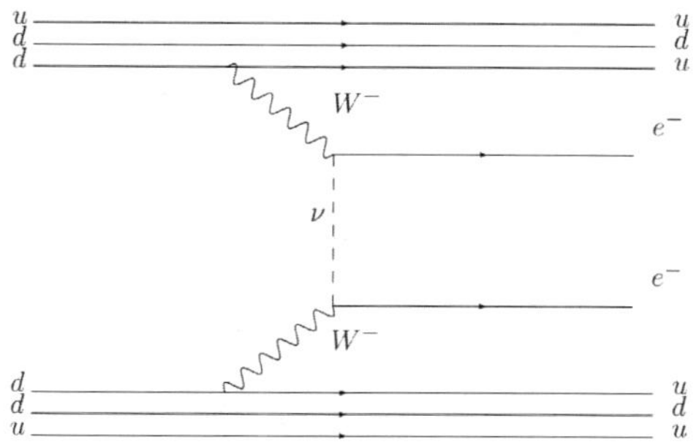

Figure 1. Feynman diagram of $0\nu\beta\beta$ decay via light Majorana neutrino exchange.

2. NEMO3

The NEMO3 detector is situated in the Modane underground laboratory (LSM) in the Frejus tunnel between France and Italy. The underground laboratory is shielded from cosmic rays by 1700 m of rock overburden, corresponding to 4800 m of water equivalent. The NEMO3 detector consists of a cylinder made of detector segments containing different samples of double beta decay enriched isotopes.

The source foils of $\beta\beta$ emitters were constructed from either a metal film or powder bound by an organic glue to mylar strips. The source hangs between two concentric cylindrical tracking volumes consisting of open octagonal drift cells operating in Geiger mode (6180 cells).

The external walls of these tracking volumes are covered by calorimeters made of large blocks of plastic scintillator (1940 blocks in total) coupled to low radioactivity PMTs. The electron tracks curve under the influence of a 25 G magnetic field to reject pair production events. The detector is substantially shielded from the external gamma ray background by 20 cm of low activity iron and 30 cm of water with boron acid to suppress the neutron flux.

There are seven different isotopes studied in NEMO3. The two isotopes with the largest mass are ^{100}Mo (6.9 kg) and ^{82}Se (0.93 kg). The other five isotopes are ^{130}Te (454g), ^{116}Cd (405 g), ^{150}Nd (37 g), ^{96}Zr (9.4 g), ^{48}Ca (7 g) which are used for $2\nu\beta\beta$ decay and background studies. NEMO3 has been taking physics data since February 2003.

The $\beta\beta$ events are selected by requiring two reconstructed electron tracks with a curvature corresponding to the negative charge, originating from a common vertex in the source foil (see Fig. 2). The detection of a clear topological signature allows unambiguous reconstruction of $\beta\beta$ events and a powerful background rejection.

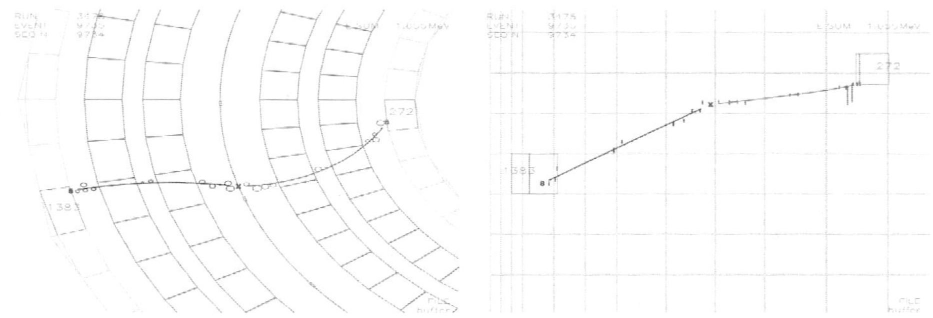

Figure 2. A $\beta\beta$ event signature observed in the NEMO3 detector. Left: top view, right: side view.

Journal of Physics: Conference Series **179** (2009) 012006 doi:10.1088/1742-6596/179/1/012006

A detailed description of the NEMO3 detector and its performance can be found in [1].

2.1. NEMO3 results.
Measurements of the $2\nu\beta\beta$ decay half-lives have been performed with unprecedented precision for the seven isotopes available in NEMO3. The results are shown in Table 1.

Table 1. NEMO-3 results of $2\nu\beta\beta$ decay half-life measurements for seven isotopes.

Isotope	Mass (g)	$Q_{\beta\beta}$ (keV)	Signal/Background	$T_{1/2}$ [10^{19} years]
^{100}Mo	6914	3034	40	0.711 ± 0.002 (stat) $\pm$ 0.054 (syst) [2]
^{82}Se	932	2995	4	9.6 ± 0.3 (stat) $\pm$ 1.0 (syst) [2]
^{116}Cd	405	2805	7.5	2.8 ± 0.1 (stat) $\pm$ 0.3 (syst) [3]
^{150}Nd	37.0	3367	2.8	$0.920^{+0.025}_{-0.022}$(stat) $\pm$ 0.072 (syst) [4]
^{96}Zr	9.4	3350	1.	2.3 ± 0.2 (stat) $\pm$ 0.3 (syst) [4]
^{48}Ca	7.0	4272	6.8	$4.4^{+0.5}_{-0.4}$(stat) $\pm$ 0.4 (syst) [4]
^{130}Te	454	2529	0.25	76 ± 15 (stat) $\pm$ 8 (syst) [3]

The most precise measurements were obtained for ^{100}Mo and ^{82}Se but the results for other isotopes are also most accurate to date. Fig. 3 shows one of the most recent results, the ^{96}Zr-isotope half-life measurement.

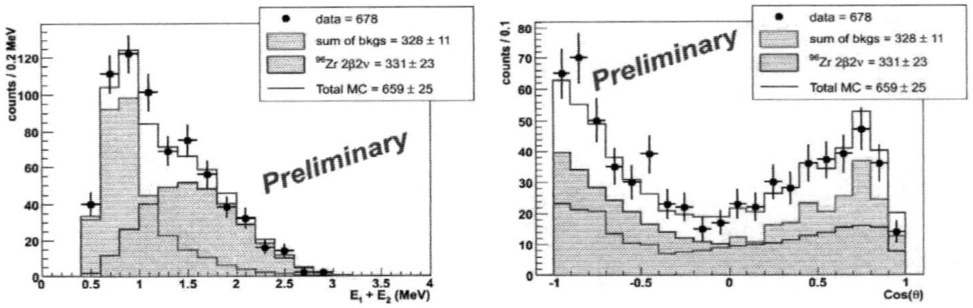

Figure 3. The sum of the energies of the two electrons and their angular separation for two-electron events for ^{96}Zr.

The distribution of the energy sums of the two electrons around the $Q_{\beta\beta}$ value was used to search for $0\nu\beta\beta$ in all isotopes. The best sensitivity was obtained with the ^{100}Mo data. Using the data until spring 2006 a limit on the $0\nu\beta\beta$ half-life of ^{100}Mo was set to be $T^{0\nu}_{1/2} > 5.8 \cdot 10^{23}$ yr at 90% CL which translates into a limit on the neutrino mass of $< m_\nu > <$ 0.8–1.3 eV. The half-life limit was also used to obtain the best constrain on Majoron-neutrino coupling to date , $\langle g_{ee} \rangle < (0.4 - 1.8) \cdot 10^{-4}$ [5]. The ranges in the values of $< m_\nu >$ and $\langle g_{ee} \rangle$ are due to uncertainties in NME calculations. The collaboration will release a new result covering the entire data taking period in summer 2009 and once again at the end of running in 2011. The final sensitivity of NEMO3 will be $< m_\nu > <$ 0.4–0.8 eV.

3. SuperNEMO

SuperNEMO aims to extend and improve the experimental techniques used by NEMO3 to search for neutrinoless double beta decay. It will extrapolate the successful NEMO3 technology over one order of magnitude by studying about 100 kg of double beta decay isotopes. The detector's ability to measure any $\beta\beta$ isotope and the topological signature are distinct features of SuperNEMO. The baseline isotope choice for SuperNEMO is ^{82}Se and ^{150}Nd , if the enrichment of the latter proves to be feasible. The experiment will reach a sensitivity to the Majorana neutrino mass, $< m_\nu >$, of 50-100 meV in 5 years of running which, according to the current schedule, will start in 2012.

3.1. SuperNEMO detector design.

The baseline SuperNEMO design envisages about twenty identical modules, each housing around 5 kg of isotope. A preliminary design of a SuperNEMO detector module is shown in Fig. 4.

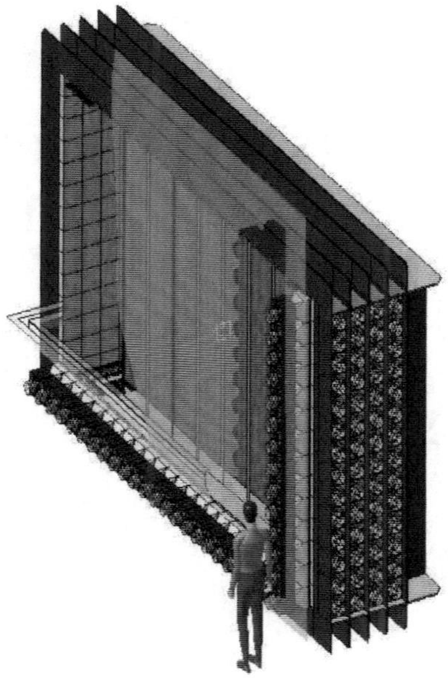

Figure 4. Schematic of a proposed SuperNEMO module showing the source foil surrounded by a tracking volume and scintillator blocks read out by PMTs.

The source is a thin ($\sim$ 40 mg/cm^2) foil inside the detector. It is surrounded by a gas tracking chamber followed by calorimeter walls. The tracking volume contains around 2000 wire drift cells operated in Geiger mode which are arranged in nine layers parallel to the foil. The calorimeter is divided into 700 plastic scintillator hexagonal blocks (25 cm diameter) which cover most of the detector outer area and are coupled to low radioactive 8-inch PMTs. In parallel with the baseline design, an alternative design is being investigated. The latter is based on long (2 m) scintillator bars read out from both ends by two 3-inch PMTs. In this case, the entire detector is built as a single module with calorimeter walls and source foils "sandwiched" between each other.

3.2. SuperNEMO design study

The SuperNEMO R&D programme (2006–2009) is currently underway, supported by funding agencies in the UK, France, Russia, Czech Republic, Spain and other countries. The programme addresses three main challenges: improvement of the calorimeter energy resolution, radio-purity of the source foils and optimisation of the tracker detector for large-scale production under low background requirements.

The expected improvement in performance of SuperNEMO compared to its predecessor NEMO3 is shown in Table 2. The most important design study tasks are described in the sections that follow.

Table 2. Comparison of the main NEMO 3 and SuperNEMO parameters.

	NEMO 3	SuperNEMO
Isotope	^{100}Mo	^{82}Se or ^{150}Nd
Mass	7 kg	100 kg
Detection efficiency	18%	$> 30\%$
^{208}Tl in foil	$< 20\mu$Bq/kg	$< 2\mu$Bq/kg
^{214}Bi in foil	$< 300\mu$Bq/kg	$< 10\mu$Bq/kg (^{82}Se)
energy resolution at 3 MeV	8% (FWHM)	4% (FWHM)
half-life	$T^{0\nu}_{1/2} > 1.5 \cdot 10^{24}$ years	$T^{0\nu}_{1/2} > 10^{26}$ years
neutrino mass	$\langle m_{\beta\beta} \rangle < 0.4 - 0.8$ eV	$\langle m_{\beta\beta} \rangle < 0.05 - 0.1$ eV

3.3. Energy resolution

An excellent energy resolution is vital for discriminating the $0\nu\beta\beta$ peak in the energy sum of the two electrons from the background of two-neutrino beta decay. The energy resolution is determined by two factors, the electron energy losses in the foil and the calorimeter resolution.

SuperNEMO aims to improve the calorimeter energy resolution to $7\%/\sqrt{E}$ at FWHM (compared to the NEMO3 resolution of $\sim 14\%/\sqrt{E}$).

A large number of studies have been carried out investigating the choice of calorimeter parameters such as scintillator material (organic plastic, liquid, or non-organic), and the shape, size and coating of calorimeter blocks. These are combined with dedicated development of PMTs with low radioactivity and high quantum efficiency. The target energy resolution (7% at 1 MeV) has been demonstrated with large hexagonal (25 cm diameter) PVT-based plastic scintillator blocks coupled to high-QE low radioactive 8" PMTs. In addition a record energy resolution has been obtained for the bar design (10% at 1 MeV). The final choice between the two designs will be made in June 2009. The next step is to demonstrate that these detector parameters can be maintained during mass production of the detector components.

3.4. Radiopurity of the source

SuperNEMO will search for a very rare process, therefore it must maintain ultra-low background levels. The source foils must be radiopure, and their contamination with naturally radioactive elements must be precisely measured. The most important source contaminants are ^{208}Tl and ^{214}Bi, whose decay energies are close to the neutrinoless signal region. In order to evaluate their activities, a dedicated BiPo detector has being developed which can measure their signature of an electron followed by a delayed alpha particle. The detector has been demonstrated to have the required sensitivity at the level of $\sim 10\ \mu$Bq/kg.

3.5. Tracker design

The SuperNEMO tracker consists of octagonal wire drift cells operated in Geiger mode. Each cell is around 4 m long and has a central anode wire surrounded by 8–12 ground wires, with cathode pickup rings at both ends. Signals can be read out from the anode and/or cathodes to determine the position at which the ionising particle crossed the cell.

The tracking detector design study looks at optimising its parameters to obtain high efficiency and resolution in measuring the trajectories of double beta decay electrons, as well as of alpha particles for the purpose of background rejection. The tracking chamber geometry is being investigated with the help of detector simulations to compare the different possible layouts. In addition, several prototypes have been built to study the drift chamber cell design and size, wire length, diameter and material, and gas mixture.

A large 90-cell prototype has been constructed and commissioned in January 2009. It will be used to finalise the tracker design and to test the mechanics and large-scale operation of the drift cell system.

A single SuperNEMO module will contain about 2000 drift cells with 8–12 wires each. The large total number of wires has lead to the need of automated wiring. A dedicated wiring robot is being developed for the mass production of drift cells.

3.6. SuperNEMO sensitivity

The sensitivity of SuperNEMO has been studied extensively during the current design study phase. A full chain of GEANT4 based simulation software has been developed and commissioned and the sensitivity was studied as a function of various detector parameters such as the calorimeter energy resolution, source foil radio-purity, tracking detector configuration, etc. A summary of these studies is shown in Fig. 5. With the target detector parameters and 500 kg-yr exposure (100 kg of ^{82}Se for 5 years of running), the calculated sensitivity is $\sim 10^{26}$ yrs at 90% CL which corresponds to an upper limit on $< m_\nu >$ of 50–110 meV depending on the NME used. This sensitivity is expected to be reached by 2017.

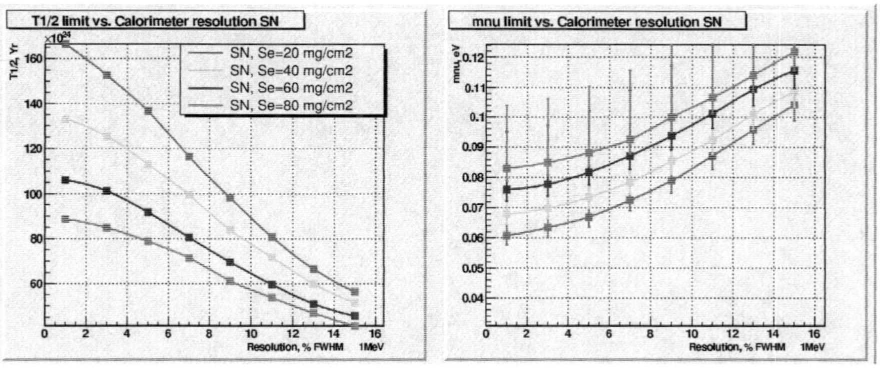

Figure 5. MC simulations of the SuperNEMO sensitivity obtained with 100 kg of ^{82}Se and 5 yr of data taking in terms of half-life (left) and Majorana neutrino mass $< m_\nu >$ (right). NME calculations from [6] are used for the $< m_\nu >$ sensitivity plot. Target values for source contamination are assumed (10 μBq/kg for ^{214}Bi and 2 μBq/kg for ^{208}Tl).

4. Summary

The NEMO approach of using tracking plus calorimetry is unique for a $\beta\beta$ experiment and allows for excellent background rejection, choice of multiple isotopes and full kinematic reconstruction leading to an observation of a “smoking gun” signature of the process. NEMO3 is taking data and have measured the $2\nu\beta\beta$ processes for several nuclei with unprecedented precision. These results provide an experimental input to nuclear matrix element calculations. In addition precise measurements of $2\nu\beta\beta$ processes are important as they provide information on this ultimate background component to $0\nu\beta\beta$. NEMO3 will run until 2011 focusing on $0\nu\beta\beta$ search and will reach a sensitivity to $< m_\nu >$ at the level of 400-800 meV.

At the same time an active R&D programme for the next-generation experiment SuperNEMO which will accommodate a 100 kg of isotope has yielded promising results. The next step is to build a first demonstrator module and install it in 2012 in LSM in place of the NEMO3 detector. The goal of this first module is to demonstrate feasibility of the target detector performance parameters, most importantly the background levels and calorimeter energy resolution. It will also produce a competitive physics measurements improving on the NEMO3 results. The rest of the modules are expected to be installed in the new-LSM underground lab (subject to availability) starting from 2013. SuperNEMO will reach a sensitivity to $< m_\nu >$ at the level of 50–100 meV by 2017.

References

[1] Arnold R *et al.* 2005 *Nucl. Instrum. Methods* A **536** 79
[2] Arnold R *et al.* 2005 *Phys. Rev. Lett.* **95** 182302
[3] Soldner-Rembold S 2008 *J.Phys conf. Ser* **110** 082019
[4] Flack R 2008 *J.Phys conf. Ser* **136** 022032
[5] Arnold R *et al.* 2006 *Nucl. Phys.* A **765** 483
[6] Rodin V *et al.* 2007 *Nucl. Phys.* A **793** 213

Energy resolution of alpha particles in a Micromegas detector at high pressure

F. J. Iguaz[1], T. Dafni[1], E. Ferrer-Ribas[2], G. Giomataris[2], P. Gordetzsky[3], I. Irastorza[1], P. Salin[3] and A. Tomás[1]

[1] Laboratorio de Física Nuclear y Astropartículas, Universidad de Zaragoza, Zaragoza, Spain
[2] Centre d'Études Nucleaires de Saclay, IRFU/CEA-Saclay, Gif-sur-Yvette, France
[3] APC-Astroparticule et Cosmologie, CNRS-CEA-IN2P3-Observatoire de Paris, Université Paris Diderot-Paris7, France

E-mail: iguaz@unizar.es

Abstract. The latest Micromesh Gas amplification Structures (Micromegas) are achieving outstanding energy resolution for low energy photons, with values as low as 11% FWHM for the 5.9 keV line of ^{55}Fe in Argon/Isobutane mixtures at atmospheric pressure. At higher energies (MeV scale), these measurements are more complicated due to the difficulty in confining the events in the chamber, although there is no fundamental reason why resolutions of 1% FWHM or below could not be reached. There is much motivation to demonstrate experimentally this fact in Xe mixtures due to the possible application of Micromegas readouts to the Double Beta Decay (DBD) search of ^{136}Xe, or in other experiments needing calorimetry and topology in the same detector. We report on systematic measurements of energy resolution with state-of-the-art Micromegas using a 5.5 MeV alpha source in high pressure Ar/Isobutane mixtures. Values as low as 1.6% FWHM have been obtained. Same measurements in Xe, of which a preliminary result is also shown here, are under progress.

1. Introduction

The neutrino flavour oscillation observed in atmospheric and solar neutrinos experiments has increased the interest on Neutrinoless DBD [1, 2]. This rare process has never been observed and its detection would give information about the mass and the Dirac or Majorana nature of the neutrino. The requirements for this kind of experiments are challenging: a large amount of DBD emitter, a selection of low-background materials, the use of signal-background discrimination techniques and good energy resolution. Time Projection Chambers (TPCs) have been generally used as tracking detectors but their energy resolution wasn't good enough to compete with Germanium or bolometers in DBD. The appearance of new charge readout planes based on micropattern techniques (like Micromegas) with better energy resolution capabilities is a further motivation for the application of these detectors in Neutrinoless DBD.

The MICROMesh GAseous Structure (MICROMEGAS)[3, 4] readout for TPCs consists of a metallic micromesh suspended over a (usually pixellised) anode plane by means of isolator pillars, defining an amplification gap of the order of 50 to 150 μm. Electrons drifting towards the readout go through the micromesh and their amplification induces signals both in the anode pixels and in the mesh. The latest generation of these detectors has obtained an energy resolution of 11% FWHM for the 5.9 keV ^{55}Fe line for a mixture of Ar - 2% iC_4H_{10} at 1 bar. Supposing

Figure 1. Upper and lower parts of the high pressure vessel. The drift cathode and the source keeper are shown at the upper part and the microbulk detector and the metallic support at the lower one.

a square root energy dependency of the energy resolution, this would lead to a value less than 1% FWHM for electrons at 2.480 MeV (the Q value of the DBD for ^{136}Xe) and would fulfill the requirements of DBD experiments [5]. The experimental confirmation of this hypothesis is the aim of this work.

In section 2 we describe the experimental setup used for the measurements. In section 3 we present the energy resolution results in Argon/Isobutane mixtures and pure Xenon. In section 4 results on the alpha quenching factor are given. We finish with our conclusions and future plans in section 5. A more detailed description of the setup and results is made in [8].

2. Setup description

The experimental setup was adapted from the former HELLAZ one at CEA-Saclay[6], which was already designed for research with gas mixtures of high purity and at high pressures. It consists of a high pressure vessel, a turbomolecular pump and a gas distribution system with a gas mixer, an oxysorb filter, a pressure controller and an exhaust line with a bubbler. The main element is a high pressure vessel of 1 liter of volume, which provides high voltage (SHV) and signal (BNC) feed-throughs for the internal readout setup, as well as a CF40 outlet for pumping, equipped with an all-metal UHV high-pressure valve. As it is shown in figure 1, the upper part of the vessel consists of a drift copper cathode, which supports an unsealed Americium source in a sourcekeeper. At the lower part, a Micromegas readout was placed on a metallic support, used to shape the drift electrical field. Both pieces were electrically isolated. The drift distance of the vessel was 31 mm and as alpha tracks in Argon and Xenon are longer, no data was taken at atmposheric pressure.

The Micromegas readout is a 2.3 cm diameter circular one of 50 microns gap. It was made with the *microbulk* manufacturing technique developed at CERN, known to yield the highest precision in the gap homogeneity and the best energy resolutions among Micropattern detectors. The one used in this measurements was tested before and after the measurements campaign with a ^{55}Fe source in Ar - 2% iC_4H_{10} at atmospheric pressure, presenting the spectrum shown in 2. The energy resolution is 11.2 % FWHM at the 5.9 keV line. No appreciable deterioration was seen after the year-long duration of the measurements campaign.

The signal was read out from the Micromegas mesh using a CANBERRA 2004 preamplifier, whose output was fed into an ORTEC VT120 amplifier/shaper and subsequently into a multichannel analyzer AMPTEK MCA-8000A for spectra building. Alternatively, the output of the preamplifier was digitized directly by a LeCroy WR6050 oscilloscope and saved into disk for an offline pulse shape analysis.

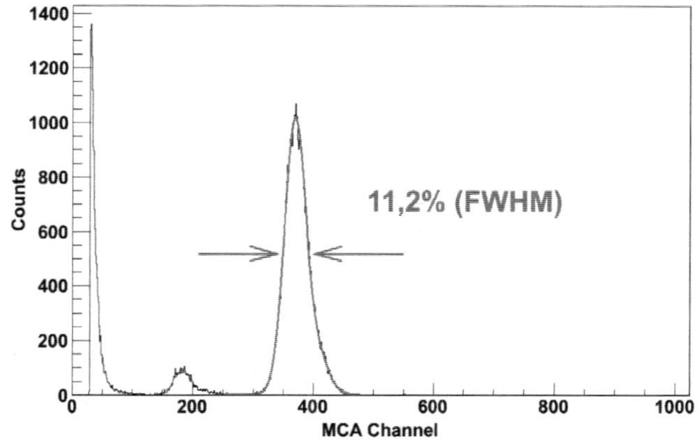

Figure 2. Energy resolution of the Micromegas readout used in the measurements. The spectrum was obtained with a ^{55}Fe source in Argon - 2 $\% iC_4H_{10}$ at 1 bar.

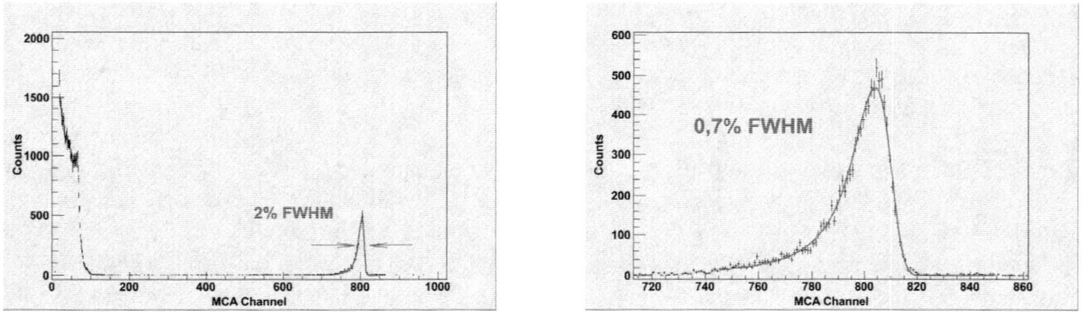

Figure 3. Americium spectrum in Argon - 2 % iC_4H_{10} at 4.75 bars. Energy resolution was better than 2% FWHM. Landau deconvolution analysis points an intrinsic 0.7% FWHM.

3. Energy resolution in Argon and Xenon

Before each measurement, the system was cleaned either by pumping it with the turbo-molecular pump down to pressures below 10^{-6} mbar or by purging it with Argon or Xenon several times. The desired gas mixture was then introduced into the vessel using the gas mixer with the appropriate relative flows. Gases used were a mixture of Argon-Isobutane (between 2% and 5%) and pure Xenon. The gas used was provided by MESSER and its purity was always 4.6 or superior. The final gas purity, determined in addition by the leak-tightness of the system, the outgassing of materials and the effect of the oxysorb, was proven to be adequate for Argon mixtures but not for pure Xenon, as evidence of some degree of attachment was found.

An energy resolution of 1.5 – 2% FWHM for mixtures of Argon-Isobutane (2 and 5%) at pressures between 2 and 5 bars was obtained, as it is shown in figure 3. This result was independent of amplification fields and electronics but a dependence on drift field was clearly observed. The peak showed an asymmetric shape which can be attributed to external effects like an uncomplete charge collection. These effects are well parameterized by an inverse Landau function convoluted by a gaussian, which contains the intrinsic energy resolution of the Micromegas. The analysis indicates an energy resolution of 0.7% FWHM.

In pure Xenon, measurements were made in pressures between 2 and 4 bars were used. The

Journal of Physics: Conference Series **179** (2009) 012007 doi:10.1088/1742-6596/179/1/012007

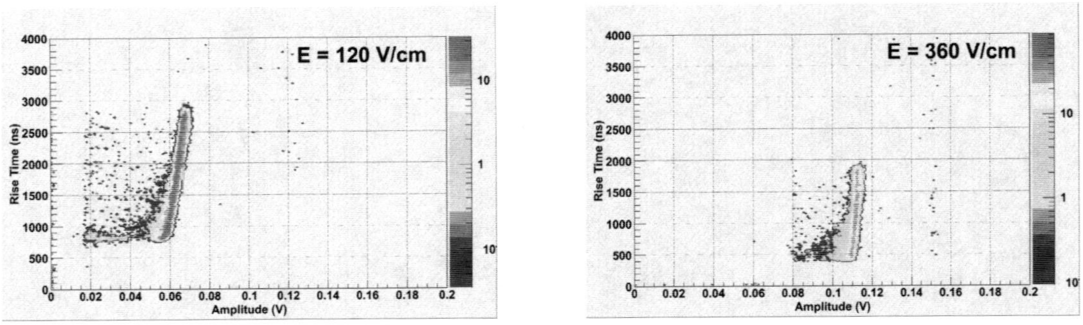

Figure 4. Rise time versus the amplitude of the signal for a drift field of 120 and 300 V/cm in pure Xenon at 2 bars.

energy resolution was not very good due to a clear presence of attachement which was not so evident in Argon-Isobutane mixtures. As it is shown at figure 4, the signal amplitude increases with the drift field until this effect is compensated with the lost of the mesh transparency. Apart from that, there is a clear dependence of the amplitude on risetime. An almost horizontal alpha track, whose electrons drift on average longer distances than those of a vertical one, has a small amplitude because its electrons suffer more the effect of the attachement. This track has also a shorter risetime due to its shorter vertical projection.

For estimating the energy resolution in absence of attachment, a cut on risetime (100-200 ns) was applied. In this way, events with similar risetime were taken in order to minimize the spread due to different attachments. Preliminary values with this analysis were 2.8% FWHM for pure Xenon at 2 bar and 4.5% FWHM at 4 bar. It should also be stressed that these numbers are conservative, because the subset of events kept by the cuts still suffers from attachment. Nevertheless, the values obtained are positive and already approach the requirements of a DBD experiment.

Another fact that should be taken into account is that absolute gains up to 200 were reached (as shown in figure 5), before getting a spark at the mesh. This value may be limited by the source rate (about 200 Hz) because higher gains were obtained with a more collimated source when trying to measure quenching factor in pure Xenon. Further measurements should be done to confirm this hypothesis.

4. Quenching factor in Argon

Alpha particles of a given energy do not ionize in theory as much as electrons of the same energy due to the electron-ion pairs recombination, which is stronger in denser alpha tracks. The fraction of energy going to ionization with respect to the one of an electron of equal energy, or *quenching factor*, is needed in order to know the electron-equivalent energy of our 5.5 MeV alpha peak. This value is difficult to find in literature for gases and it may depend on pressure, electric field and the presence of quenchers.

The x-rays and gammas can be easily observed when blocking alphas with an aluminium foil (figure 6, left). However as it is needed to keep the same gas conditions, the source was collimated to reduce alpha-induced background and sparks generated at higher gains. The same lines were observed (figure 6, right) but without increasing too much the alpha background level at the low energy region.

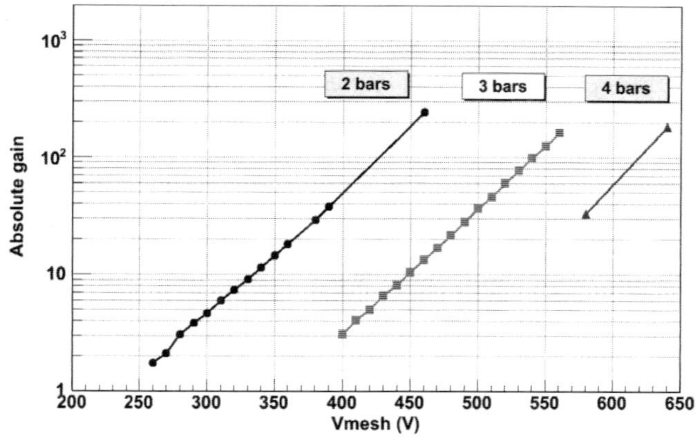

Figure 5. Absolute gain in pure Xenon versus mesh voltage for pressures between 2 and 4 bars.

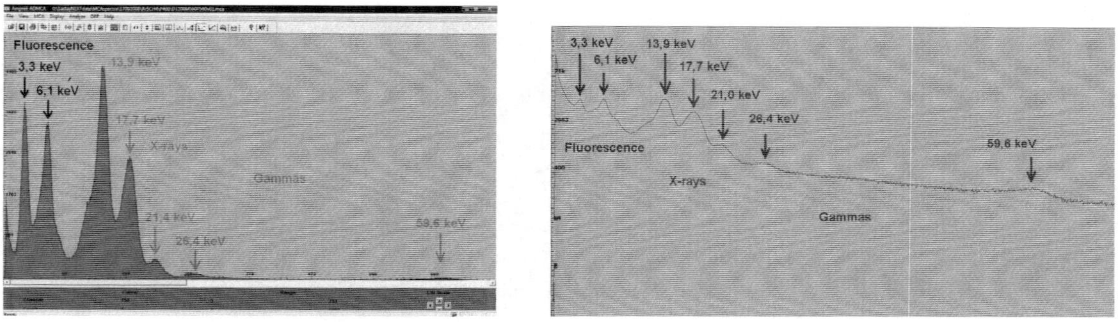

Figure 6. Americium x-rays and gamma lines blocking alphas with an aluminium foil (left) and collimating the Americium source (right).

Gain curves for both the alpha and the gamma peaks were measured for Ar - 2% iC_4H_{10} at pressures of 2, 3 and 4 bars. The quenching factor was almost unity in all cases. Similar results were obtained with Ar - CH_4 mixtures. More details are given at [8].

5. Outlook and conclusions

We have presented first results from measurements of energy resolution at high energy and high pressures for a Micromegas readout. A value of less than 2% (FWHM) in Argon-Isobutane mixtures ($2-5\%$) at pressures between 2 and 5 bars was obtained. An indication of an intrinsic resolution of 0.7% based on the asymmetry of the peak was observed. Measurements in pure Xenon were made. By performing a pulse analysis on the digitized waveforms of the preamplifier output and cuts on the risetime, a conservative estimation of the energy resolution of 2.8% for 2 bar and 4.5% for 4 bar was found. An alpha quenching factor of almost unity was obtained for Ar - 2% iC_4H_{10} at 2, 3 and 4 bars. A more detailed description of the setup and results is made in [8].

This is the first time that energy resolution measurements of alpha particles at high pressure with microbulk type of Micromegas are published. Similar measurements were done in [7] although with considerable worse results. At present we are working on the upgrade of the system to improve its leak-tightness and outgassing in order to resume the measurements in pure

Xenon in appropriate conditions of gas purity. We also plan to use an enlarged high pressure vessel big enough to contain high energy electron events, and perform similar measurements with photon/electron sources and not only with alphas. Finally, we plan to explore the effect of quenchers in Xenon in the energy resolution as well as going up to higher pressures than the ones in the present paper (from 5 to 10 bars).

Acknowledgments

This work was supported by the Spanish Science and Innovation Ministery (FPA2004-00973 and FPA2007-62833) and a grant of the same institution (AP2005-0360).

We would also like to thank all the technical support of the SEDI departement at the IRFU/CEA-Saclay center.

References

[1] Angel Morales, *Nucl. Phys. (Proc. Suppl.)* **77** (1999), 335-345.
[2] S.R. Elliot and J. Engel, *J. Phys G* **30** (2004), R183.
[3] Y. Giomataris, P. Rebourgeard, J. P. Robert and G. Charpak, *Nucl. Instrum. Meth. A* **376** (1996) 29.
[4] I. Giomataris *et al.*, *Nucl. Instrum. Meth. A* **560** (2006) 405, *Preprint* arXiv:physics/0501003.
[5] F. T. Avignone, G. S. King and Yu. G. Zdesenko, *New J. Phys.* **7** (2005) 6.
[6] J. Dolbeau, I. Giomataris, P. Gorodetzky, T. Patzak, P. Salin and A. Sarat, *Nucl. Phys. Proc. Suppl.* **138** (2005) 94.
[7] L. Ounalli, J. L. Vuilleumier, D. Schenker and J. M. Vuilleumier, *JINST* **4** (2009) P01001, *Preprint* arXiv:hep-ex/0810.0445.
[8] Th. Dafni et al., *Preprint* arXiv:0906.0534.

Journal of Physics: Conference Series **179** (2009) 012008 doi:10.1088/1742-6596/179/1/012008

New Results of GridPix TPCs

Harry van der Graaf

Science Park 105
1098 XG Amsterdam, The Netherlands

vdgraaf@nikhef.nl

Abstract. The Gossip detector, being a GridPix TPC equipped with a thin layer of gas, is a promising alternative for Si tracking detectors. In addition, GridPix would be an interesting way to read out the gaseous phase volume of bi-phase Liquid Xe cryostats of ν-less double beta decay and rare event (i.e. WIMP) search experiments.

1. The GridPix detector

100 years ago, in Manchester, Hans Geiger operated the first gaseous detector, which was the basis for 'wire chambers', widely applied as track detectors in particle physics experiments. In wire chambers gas amplification occurs, close to the wire surface, due to the strong (1/R) electric field. This enables the detection of the few single electrons created in the gas by ionization radiation. In Micro Pattern Gas Detectors, areas with a strong electric avalanche field are created by one or more conductive perforated (grid) planes. The granularity of such a detector is determined by the pitch of the holes and can be much better in comparison to wire chambers. With each grid hole equipped with its own readout channel (preamp, shaper and discriminator) in the form of an active pixel array in a CMOS chip, this micro-granularity is pursued: each hole is a stand-alone detector [1]. This GridPix detector matches future demands on occupancy, position resolution and time resolution for high radiation trackers at future ILC, CLIC or sLHC colliders. Chip manufacturing processes made two innovations possible: the integration of Micro Pattern Gas Detectors with pixels chips (Integrated Grid, InGrid), and the deposit of a high-resistivity protection layer on top of pixel chips. This 'wafer post processing' technology may enable next innovations such as micro channel plates and secondary emission foils. Essentially, the application of gas as detection material, compared to, for instance, Si, offers several advantages, relevant for future tracking and imaging detector developments. For this, the development of TimePix-2, a new general-purpose pixel chip, is essential.

1. 1 The Integrated Grid (InGrid)

We constructed a Micromegas, by means of 'Wafer Post Processing' technology, on top of de MediPix-2 and TimePix pixel chips [2]. The inherent precision of this technology allows a perfect alignment of the grid holes with the pixel input pads, sub-micron precision in the grid hole diameters and a constant thickness of the avalanche gap; see fig.1. We have recorded an unprecedented energy resolution (5.2 % r.m.s) of the photopeak of ^{55}Fe pulses in proportional gaseous detectors [3]. The technology, applied on single chips, is being prepared to be applied on wafers.

The present material used for InGrid is aluminum. In a new development we would like to use high-resistivity material for this, like amorphous silicon. This would add to the discharge quenching power.

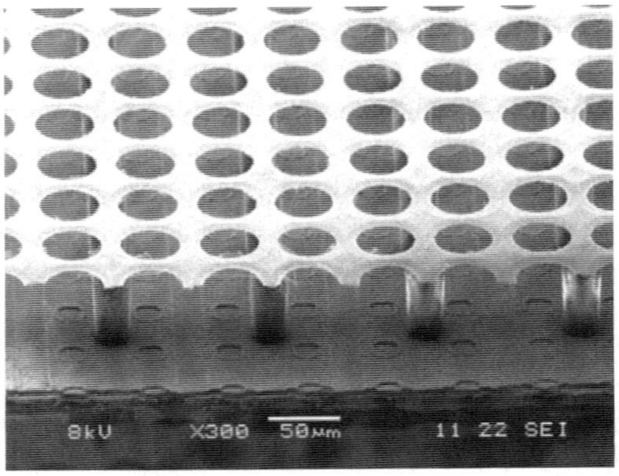

Figure 1. The Integrated Grid (InGrid) on top of a 20 μm SiProt layer, on top of a TimePix chip.

1.2 Gossip: gas versus Silicon

GridPix can be considered as an active readout system for a gaseous volume in which an electric (drift) field is applied. The maximum drift length, equal to the 'thickness' of the gas volume, may vary between 2 m in the case of large TPCs, and 1 mm in case of the Gossip detector [4]. This detector is an alternative for Si pixel- and strip detectors where the Si is replaced by a thin gas layer. We have operated a Gossip detector which was based on the CMS pixel FE chip PSI-46.

Gas, as detection material for tracking detectors, has several advantages with respect to Si:

- gas can be exchanged: radiation damage does not occur;
- by applying a strong electric field, noiseless gas amplification is possible, resulting in arbitrary large charge signals. This reduces the required power for the input circuits, and thus the required materials for cooling;
- the source capacitance, seen at the input of preamplifiers, is minimal;
- gas is light;
- gas is generally cheap;
- gas detectors are little sensitive for (background) X-rays, gammas and neutrons;
- there is no bias current in gaseous detectors;
- gaseous detectors do not require to be operated in a cooled environment;
- δ-rays can be recognized and do not cause an error in the track position measurement.

Gaseous detectors have two disadvantages: discharges occur, potentially destroying (parts of) the detector, and chamber ageing. The discharge issue has been solved [5]; the knowledge on chamber ageing has recently increased dramatically [6].

2. The application of GridPix in bi-phase LXe rare event detectors

With the gaseous planar GridPix detector, the (2D) position and the time of arrival can be registered of electrons, after their drift from their point of creation. The efficiency of detecting single electrons can be as high as 99 percent, while the noise, appearing solely as falsely detected single electrons, can be extremely low. This GridPix detector would lower the threshold of detecting rare (WIMP) events, or would improve the energy resolution in 0vee experiments. The measured ionisation pattern would provide information on the direction of a WIMP interaction, or, in 0vee experiments, the electron energy or the distribution over both electrons.

2.1 WIMP signals in liquids or gases

A recoil of a nucleus creates a track of electron-ion pairs, with a threshold of one electron. After their drift, these electrons can be detected by the GridPix readout system. If the time of creation is measured by means of PMs, the position, in three dimensions, of all participating electrons can be obtained, and therefore the ionisation pattern. The pattern and total number of electrons define the possible interaction. A large number of participating electrons could be due to neutron capture or gamma absorption. A small number of participating electrons can be due to X-ray absorption, but this can be excluded if there is no nearby ionisation: the smaller number of participating electrons, the littler the X-ray energy and range. The fiducial volume of the experiment is related to the amplitude of the events (self-shielding). The extreme threshold can be reached if the readout is sensitive for single electrons.

Assuming a cylindrical volume with vertical axis, PMs could be placed at the bottom plane and cylinder surface. The top (circular) surface could be covered with a continuous Gridpix surface, consisting of TimePix chips equipped with InGrid. GridPix would be sensitive for arriving electrons continuously, untriggered. Triggers from PM tubes would results in drift time-correlated electron measurement.

The measurement of the noise in GridPix is essential and well possible by switching off the drift field. This experiment does neither require to be placed in a mine to avoid cosmic muons, nor needs to be made of low-background material. These background events are well recognised, and, in fact, contribute to the knowledge of the (pixel) detector.

References

[1] H. van der Graaf: GridPix: An integrated readout system for gaseous detectors with a pixel chip as anode. Nucl. Instr. Meth. A 580 (2007) 1023
[2] M. Campbell et al., Nucl. Instr. and Methods A 540 (2005) 295 - 304
[3] M. Chefdeville et al.: Nucl. Instr. and Methods A 591 (2008) 147
[4] H. van der Graaf et al.: GOSSIP: A vertex detector combining a thin gas layer as signal generator with a CMOS readout pixel array. Nuclear Instruments and Methods in Physics Research Section A: Accelerators, Spectrometers, Detectors and Associated Equipment, Volume 560, Issue 1, 1 May 2006, Pages 131-134
[5] Discharge protection layers for CMOS pixel anode chips in MPGD-GridPix detectors. To be published in Nucl. Instr. & Methods.
[6] Analysis of Compounds Released From Various Detector Materials and Their Inpact on Ageing Characteristics of Proportional Counters. K. Kurvinen et al., IEEE Transactions on Nuclear Science, Vol 51, No 5, Oct. 2004.

The DMTPC project

Gabriella Sciolla

Massachusetts Institute of Technology, 77 Massachusetts Ave., Cambridge, MA 02139, USA

E-mail: sciolla@mit.edu

for the DMTPC Collaboration

J. Battat, T. Caldwell, D. Dujmic, P. Fisher, S. Henderson, R. Lanza, A. Lee, J. Lopez, A. Kaboth, G. Kohse, J. Monroe, T. Sahin, G. Sciolla, R. Yamamoto, H. Yegoryan (Massachusetts Institute of Technology); S. Ahlen, K. Otis, H. Tomita (Boston University); A. Dushkin, H. Wellenstein (Brandeis University).

Abstract. The DMTPC detector is a low-pressure CF_4 TPC with optical readout for directional detection of Dark Matter. The combination of the energy and directional tracking information allows for an efficient suppression of all backgrounds. The choice of gas (CF_4) makes this detector particularly sensitive to spin-dependent interactions.

1. Directional Dark Matter detection

The goal of a directional Dark Matter detector [1] is to provide an unambiguous observation of Dark Matter (DM) by measuring both the direction and the energy of the nuclear recoil produced by the interaction of a Weakly Interacting Massive Particle (WIMP) with a nucleus in the active mass of the detector.

The simplest models of the distribution of WIMPs in our Galaxy suggest that the orbital motion of the Sun about the Galactic center will result in an Earth-bound observer to experience a WIMP wind with speed 220 km/s (the galacto-centric velocity of the Sun) originating from the direction of the Sun's motion. Because the Earth's rotation axis is oriented at 48° with respect to the direction of the WIMP wind, the average direction of DM particles recorded by an Earth-bound observer changes by about 90° every 12 hours [2]. The ability to measure such a direction would provide a powerful suppression of insidious backgrounds such as cosmogenic neutrons and neutrinos from the Sun [3], as well as a unique instrument to test local DM halo models [4, 5]. This capability makes directional detectors unique observatories for underground WIMP astronomy.

Dark Matter interactions in the detector generate nuclear recoils with typical energies of few tens of keV (Fig. 1, left). The direction of the recoiling nucleus encodes the direction of the incoming DM particle (Fig. 1, right). To observe the daily modulation in the direction of the DM wind, an angular resolution of $\approx 30^\circ$ in the reconstruction of the recoil nucleus is required.

Directional DM experiments use low-pressure (40–100 torr) gas as target material, in which typical DM induced nuclear recoils have a length of a few millimeters. A 3-D reconstruction of the recoil track with a spatial resolution of several hundred microns in all three coordinates is sufficient to achieve the desired angular resolution. A 2-D reconstruction is still valuable,

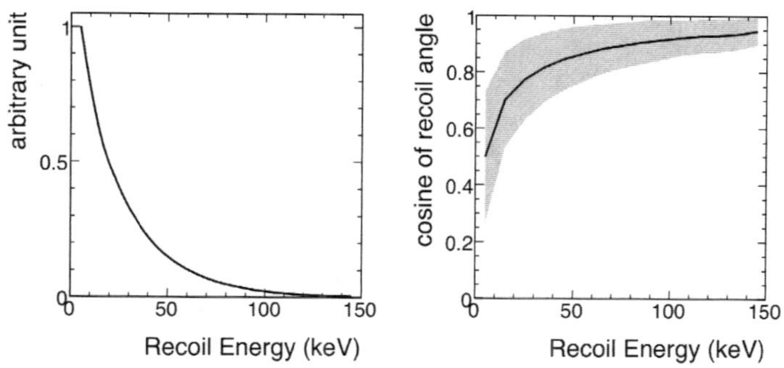

Figure 1. Energy spectrum and distribution of the cosine of the recoil angle vs. recoil energy for fluorine recoils induced by 100 GeV WIMPs in CF_4. The recoil angle is defined as the angle between the direction of the recoil and the direction of the incoming particle.

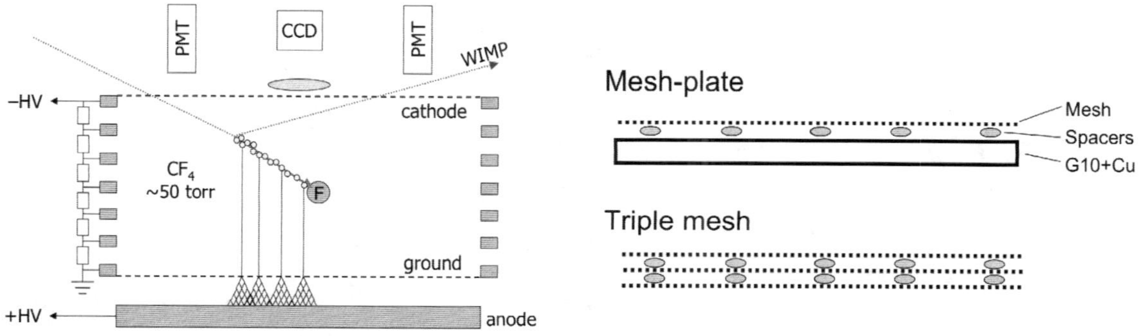

Figure 2. Left: illustration of the DMTPC detector concept. Right: two designs for the amplification region: "mesh-plate" (top) and "triple mesh" (bottom).

although it slightly degrades the sensitivity [6]. "Head-tail" discrimination is also very important as it improves the sensitivity for detecting DM by about one order of magnitude [6].

Due to the low-density of the target material, directional detectors tend to be large in volume: a typical directional detector with a fiducial mass of one ton occupies $O(10^3)$ cubic meters. Therefore, the success of a directional DM program depends very strongly on the developement of detectors with a low cost per unit volume. The largest expense for standard gaseous detectors is the cost of the readout electronics, making it essential to utilize low-cost readouts in order to make DM directional detectors financially viable.

2. The DMTPC detector concept

The DMTPC detector is a low-pressure TPC with optical readout. The detection principle is illustrated in Fig. 2, left. The TPC is filled with CF_4 at a pressure of about 50 torr. At this pressure, a typical 50 keV nuclear recoil has a length of about 2 mm. The average ionization energy in CF_4 is 54 eV [7], which results in about 10^3 primary electrons from the nuclear recoil. Such electrons drift toward the amplification region, where photons are produced together with electrons in the avalanche process. A CCD camera mounted above the cathode mesh records the image of the nuclear recoil projected along the amplification plane. The sense of the recoiling nucleus can be determined ("head-tail" discrimination), because the energy loss (dE/dx) is not constant along the trajectory. An array of photomultipliers (PMTs) mounted above the cathode mesh determines the length of the recoil in the drift direction through time measurements.

CF_4 was chosen as the target material primarily because of its high fluorine content. Fluorine is an excellent element to detect spin-dependent interactions on protons [8], due to its large spin factor and isotopic abundance (Table 1). Spin-dependent (SD) interactions are predicted to dominate over spin-independent ones in models where the lightest super-symmetric particle has a substantial Higgsino contribution [9, 10].

CF_4 is also an excellent detector material [11, 12]. Its good scintillation properties are very important for the optical readout. Our recent measurements [13] indicate that in CF_4 the number of photons produced between 200 and 800 nm wavelength is about 1 for every 3 avalanche electrons. Moreover, the low transverse diffusion characteristic of electrons in CF_4 allows for good spatial resolution in the reconstruction of the recoil track despite the long (25 cm) electron drift distance. Finally, CF_4 is non-flammable and non-toxic allowing for safe operation underground.

CCDs provide a true 2-D readout at a much lower cost-per-channel than any other readout technology used in particle physics. A modern low-noise CCD camera with 4 megapixels can be purchased today for a few thousand US dollars, which corresponds to 10^{-3} dollars/channel. The cost of a directional DM detector is dominated by the readout electronics, making the optical readout a solution toward an economically viable detector.

Two alternative implementations of the amplification region [14] are illustrated in Fig. 2. In the "mesh-plate" design (top) the amplification is obtained by applying a large potential difference ($\Delta V = 0.6$–1.1 kV) between a copper plate and a conductive woven mesh. A uniform distance of 0.5 mm between the plate and the mesh is ensured by the use of fishing wires spaced 2 cm apart. The copper or stainless steel mesh is made of 28 μm wire with a pitch of 256 μm. The pitch of the mesh determines the intrinsic spatial resolution of the detector. The "triple mesh" design (bottom) yields a transparent amplification region, which enables a single CCD camera to image two drift regions, leading to a substantial cost reduction.

The DMTPC detector measures the number of photons in the CCD camera, the 2-D image of the recoiling nucleus, the distribution of energy loss along the recoil track, the width and integral of the PMT signal, and the electronic signal produced on the amplification plane.

The energy of the nuclear recoil is independently determined by the measurement of the number of photons observed in the CCDs, the integral of the electronic signal produced on the amplification mesh, and the integral of the PMT signal. The redundancy in the design is intentional and has the goal of maximizing the robustness of the measurement. The track length and direction of the recoiling nucleus are reconstructed by combining the measurement

Table 1. Spin of the nucleus J, nuclear spin factor $\lambda^2 J(J+1)$, and abundance in nature of various isotopes considered for SD-interaction studies. The figure of merit is defined as the product of the square of the nuclear mass, the number of isotopes per kg, and the spin factor.

Isotope	J	λ^2J(J+1)	Natural abundance	Figure of merit
^{19}F	1/2	0.65	100%	74
^{23}Na	3/2	0.04	100%	6
^{73}Ge	9/2	0.06	7.8%	2
^{93}Nb	9/2	0.16	100%	91
^{127}I	5/2	0.01	100%	5
^{129}Xe	1/2	0.12	26%	25

of the projection along the amplification plane (from pattern recognition in the CCD) and the projection along the direction of drift (from the width of the signal recorded in the PMTs). The sense of the recoil track can be determined by the Bragg curve. The PMTs will also provide a measurement of the absolute z position of the recoil by measuring the time between the primary ionization and the amplified signal. This information will drastically reduce possible backgrounds from the amplification meshes.

The CCD images have a long (1–10 seconds) exposure. If during the exposure a trigger is generated by the PMT or the electronic readout of the amplification plane, the CCD is read out and the event is saved to disk. Otherwise, the CCD is reset, minimizing dead time.

The combination of the measurements described above is very effective in reconstructing the energy, direction, and sense of nuclear recoils from WIMPs. In addition, an excellent rejection of the e^- and α backgrounds can be obtained by combining the measurement of the energy and length of the recoil (see below).

3. R&D results: demonstration of directional detection

An ^{241}Am source that produces 5.5 MeV alpha particles is used as calibration source. The image of one alpha particle stopped inside CF_4 at 100 torr is shown in Fig. 3, left. Fig. 3, right, shows the Bragg curve for 5.5 MeV alpha particle in CF_4 at 280 torr. The histogram shows the energy loss measured in the prototype, while the solid line shows the results of a SRIM [15] simulation. The agreement between data and MC is better than 10%. The alpha source is used to study the gain of the detector as a function of the voltage in the amplification region and gas pressure, while a ^{55}Fe can be used for absolute energy calibration. The alpha source is also used to measure the resolution as a function of the drift distance of the primary electrons to quantify the effect of the diffusion. Our studies [16] show that the transverse diffusion has $\sigma < 1$ mm for a drift distance of 25 cm.

The first demonstration of the DMTPC detector concept was obtained with a small chamber that used parallel wire planes to obtain charge amplification. This detector was used to perform the first observation of the "head-tail" effect in nuclear recoils generated by low-energy neutrons [16]. The determination of the sense and direction of nuclear recoils was evaluated by studying the recoil of fluorine nuclei in interactions with low-energy neutrons. For nuclear recoils below 1 MeV only the tail of the Bragg peak is visible, and therefore the energy deposition decreases along the path of the recoil, allowing for the identification of the "head" ("tail") of the event by a smaller (larger) energy deposition. This initial measurement used 14 MeV neutrons

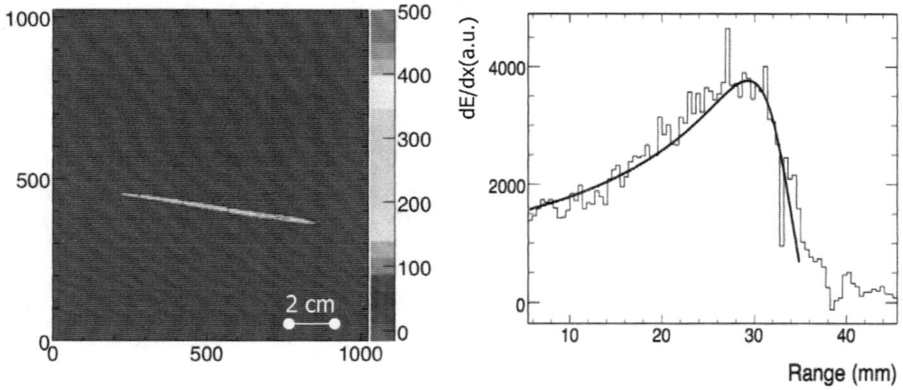

Figure 3. Left: an α track stopped in CF_4 at 100 torr (direction: left to right). Right: average light yield vs. range of the α tracks in data (histogram) compared to the SRIM MC prediction [15].

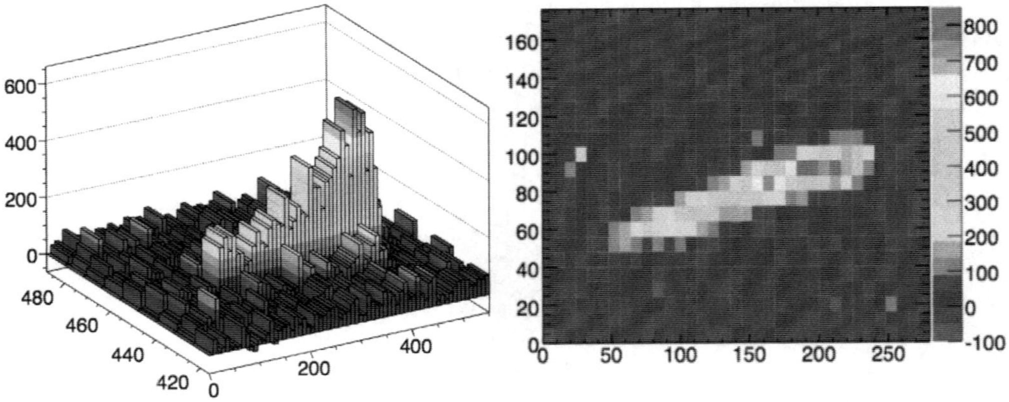

Figure 4. Left: nuclear recoil from a 14 MeV neutron in CF_4 at 250 torr recorded with the wire-based DMTPC prototype. Right: nuclear recoil induced by a neutron from a ^{252}Cf source in CF_4 at 75 torr. In both cases the neutrons were traveling right to left. The higher dE/dx visible on the right of the track is consistent with observation of the "head-tail" effect.

from a D-T generator. The reconstructed recoils had energy between 200 and 800 keV. The energy distribution along the recoil track for one of such recoils is shown in Fig. 4, left. We measured that $(73 \pm 3)\%$ of the recoils had the correct sense. This measurement represents an observation of the the "head-tail" effect with a significance of 8 σ [16].

A newer mesh-based detector was later used to study the "head-tail" effect in lower energy neutrons generated by a ^{252}Cf source. This detector allowed us to obtain a 2-D reconstruction of the nuclear recoils. Better sensitivity to lower energy thresholds was achieved by lowering the CF_4 pressure to 75 torr. Fig. 4, right shows a Cf-induced nuclear recoil due to a neutron traveling right to left inside our detector. The decreasing dE/dx along the track direction clearly visible in the image proves that the detector is able to determine the sense of the direction on an event-by-event basis. Measurements of the track length as a function of the recoil energy and of the recoil angle are shown in Fig. 5. The data (black dots) are in good agreement with the predictions of the SRIM [15] MC (histogram). Our recent measurements [14] demonstrated "head-tail" discrimination down to 100 keV at 75 torr. Monte Carlo studies indicate that this effect can be observed for recoils above 50 keV at a pressure of 50 torr when using Apogee U6 CCD cameras.

The current DMTPC prototype consists of two optically independent regions contained in

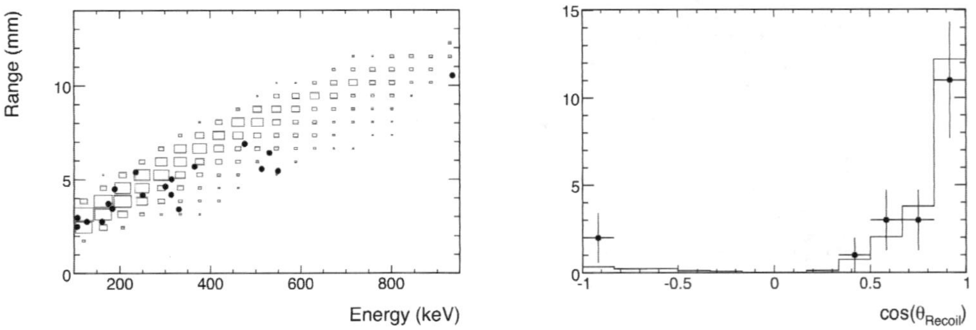

Figure 5. Range vs. reconstructed energy (left) and signed cosine of the recoil angle (right) of nuclear recoil candidates from ^{252}Cf at 75 torr. Black points are data; histogram is simulation.

one stainless steel vessel. Each region is a cylinder with 25 cm diameter and 25 cm height. A field cage made of stainless steel rings keeps the uniformity of the electric field within 1% in the fiducial volume. The amplification is obtained by using a "mesh-plate" design (Fig. 2, top-right). The detector is read out by two CCD cameras, each imaging one drift region. The optical system consists of two Nikon photographic lenses with f-number of 1.2 and focal length of 55 mm, and two Apogee U6 CCD cameras [17] equipped with Kodak 1001E CCD chips. Because the total area imaged is 16×16 cm^2, the detector has an active volume of about 10 liters.

4. DMTPCino detector design

DMTPCino is a 1-m^3 DMTPC detector designed to substantially improve limits on spin-dependent interactions of WIMPs on protons while testing the DMTPC detector concept on a larger scale and in a realistic environment. The preliminary design of the detector is shown in Fig. 6. The apparatus consists of a stainless steel vessel of 1.3 m diameter and 1.2 m height. Nine CCD cameras and nine PMTs are mounted on each of the top and bottom plates of the vessel, separated from the active volume of the detector by an acrylic window. The detector consists of two optically separated regions. Each region is equipped with a triple-mesh amplification device, mounted in between two symmetric drift regions. Each drift region has a diameter of 1.2 m and a height of 25 cm, for a total active volume of 1 m^3. A field cage keeps the uniformity of the electric field within 1% in the fiducial volume. A gas system circulates and purifies the CF_4.

All materials used inside the active volume of the detectors are selected to be radio-pure to limit backgrounds from internal radioactivity. Pure copper, stainless steel, Ph bronze, Vesper, quartz and acrylic are known to satisfy these requirements [18]. Because all CCD cameras, lenses, and PMTs are outside the active volume, their contribution to internal radioactivity is negligible.

At a pressure of 50 torr and 21 degrees C, this module will contain 250 g of CF_4. Assuming an overall data-taking efficiency of 50%, a one-year run will yield an exposure of 45 kg-days.

The CCD cameras we plan to use for DMTPCino are the same Apogee Alta U6 that we are currently using on the 10-liter detector. The U6 uses a Kodak Grade 2 KAF-1001E (1024×1024 pixel array, each pixel 24×24 μm^2) full-frame sensor with peak quantum efficiency of 72% at 560 nm, and a readout noise of 13 (20 max) electrons.

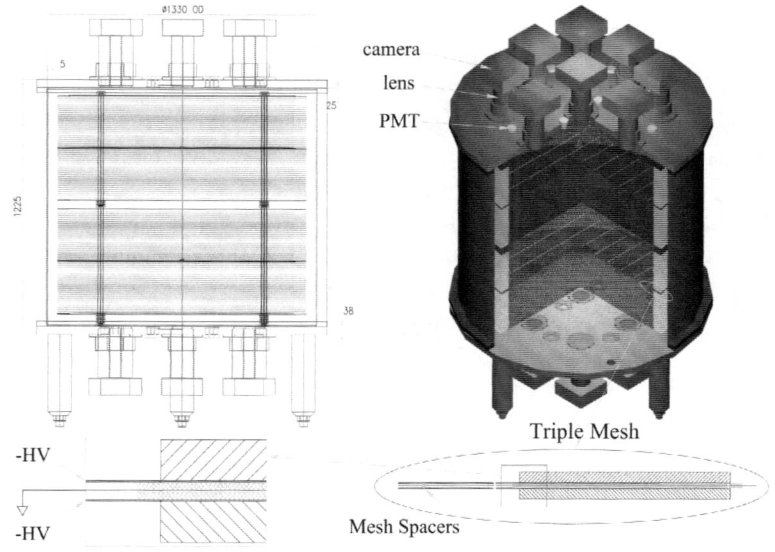

Figure 6. Preliminary drawings of the 1 m^3 DMTPC detector.

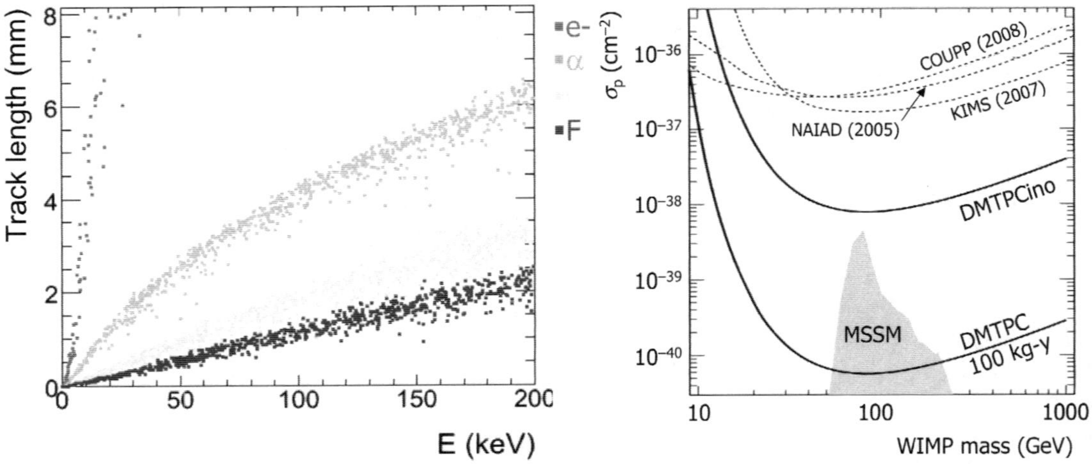

Figure 7. Left: range versus energy of the recoil track for fluorine recoils (blue), carbon recoils (yellow), alphas (green), and electrons (red). Right: expected 90% C.L. limits on SD interactions of WIMPs on protons for DMTPCino (top solid line) and a larger DMTPC detector (bottom solid line). The dashed lines show the best present limits. The blue area shows the MSSM expectation.

5. Background suppression

The main source of background for a DMTPC detector, as for any DM experiment, is due to electromagnetically interacting particles such as photons and electrons, and to α particles. These particles are produced by natural radioactivity of the detector components and surrounding materials, as well as by cosmic rays. The impact of these backgrounds will be substantially suppressed by using only radio-pure materials in the fabrication of the apparatus and by shielding the detector. In addition, the directional information provides a powerful signature that can be used to distinguish particle species: the correlation between the energy and length of the recoil track.

Fig. 7, left, shows the distribution of the length versus energy of the track expected in the DMTPC detector for electrons (red), alpha particles (green), and carbon (yellow) and fluorine (blue) nuclear recoils from WIMP interactions. The MC simulation used is known to reproduce current data within 10%. From this plot, we conclude that an excellent discrimination against α particles can be achieved with a threshold of 30 keV or lower. The gamma ray rejection factor, measured using a ^{137}Cs source, is better than 2 parts per million.

Neutrons backgrounds will be reduced by operating the detector underground and making use of a passive shielding. Residual neutrons will be rejected using the directional information.

Directionality is also key to distinguish between WIMP signal and coherent scattering of solar neutrinos, which will point back to the sun.

6. Expected sensitivity from DMTPCino

The sensitivity of the DMTPCino detector to SD interactions of WIMPs on protons has been studied assuming that the detector will be operated at a threshold of 30 keV[1] in an underground laboratory at a depth $<$ 1,600 m.w.e., and that it will be surrounded by a neutron shielding consisting of 40 cm of polyethylene. No significant internal backgrounds are assumed to be present above threshold. The limits expected from DMTPCino (5 months of data-taking at

[1] Assuming a quenching factor of about 55% for recoils of 50 keV, as predicted in reference [19], a 30 keV threshold in the detector corresponds to a nuclear recoils with kinetic energy of 50 keV.

50% efficiency) are shown in Fig. 7, right. Improvements of a factor of 50 over the best existing measurements [20, 21, 22] can be obtained despite the limited target mass due to the large SD cross-section of fluorine (Table 1) and to the excellent background rejection capability of the detector.

A large DMTPC detector, with an active mass of 10^2–10^3 kg, will be able to explore a significant portion of the Minimal Supersymmetric Standard Model (MSSM) parameter space [10] (Fig. 7, right). Such a detector is an ideal candidate for the DUSEL laboratory.

7. Conclusion

The DMTPC detector is designed to measure the energy, position, direction, and sense of low-energy nuclear recoils generated by a WIMP interaction. The combination of the energy and tracking information allows for an efficient suppression not only of backgrounds due to electrons, photons, and alphas, but also of more insidious backgrounds, such as neutrons and neutrinos. The choice of gas (CF_4) makes this detector particularly sensitive to spin-dependent interactions. We estimate that with a 1-m^3 detector (DMTPCino) in less than one year of data-taking underground we will be able to improve the existing limits on SD interactions on protons by about a factor 50. Such a detector will lay the foundation for the design of a large directional DM detector that could provide an unambiguous positive observation of WIMPs as well as a way to discriminate between the various galactic DM halo models.

8. Acknowledgments

This work was supported by the Advanced Detector Research Program of the U.S. Department of Energy, the National Science Foundation, the Pappalardo Fellowship program, the MIT Kavli Institute for Astrophysics and Space Research, and the Physics Department at the Massachusetts Institute of Technology.

References

[1] G. Sciolla, arXiv:0811.2764 [astro-ph] (Submitted to Mod. Phys. Lett. A).
[2] D. N. Spergel, Phys. Rev. D **37**, 1353, (1988).
[3] J. Monroe and P. Fisher, Phys. Rev. D **76**, 033007, 2007.
[4] M. S. Alenazi and P. Gondolo, Phys. Rev. D **77**, 043532 (2008).
[5] B. Morgan, A. M. Green and N. J. C. Spooner, Phys. Rev. D **71**, 103507 (2005).
[6] A. Green and B. Morgan, Astropart. Phys. **27**, 142 (2007).
[7] T. Tanimori, H. Kubo, K. Miuchi, T. Nagayoshi, R. Orito, A. Takada and A. Takeda, Phys. Lett. B **578**, 241 (2004).
[8] R. J. Ellis and R. A. Flores, Phys. Lett. B **263**, 259 (1991).
[9] U. Chattopadhyay and D. P. Roy, Phys. Rev. D **68**, 033010 (2003); B. Murakami and J. D. Wells, Phys. Rev. D **64**, 15001 (2001); J. Vergados, J. Phys. G **30**, 1127 (2004).
[10] J. R. Ellis, A. Ferstl and K. A. Olive, Phys. Rev. D **63**, 065016 (2001).
[11] A. Pansky *et al.*, Nucl. Instrum. Meth. A **354**, 262 (1995).
[12] L. G. Christophorou, et. al., J. Phys. Chem. Ref. Data **25**, 1341 (1996)
[13] A. Kaboth *et al.*, [DMTPC Collaboration], Nucl. Instrum. Meth. A **592**, 63 (2008).
[14] D. Dujmic *et al.*, [DMTPC Collaboration], Astropart. Phys. **30**, 58 (2008).
[15] J. F. Ziegler *et al.*, Pergamon Press, New York, 1985.
[16] D. Dujmic *et al.*, [DMTPC Collaboration], Nucl. Instrum. Meth. A **584** 327 (2008).
[17] See Apogee web site: www.ccd.com.
[18] D. S. Leonard *et al.* [EXO Collaboration], Nucl. Instrum. Meth. A **591**, 490 (2008).
[19] A. Hitachi, arXiv:0804.1191v1 [astro-ph], Submitted to Rad. Phys. Chem.
[20] H. S. Lee. *et al.* [KIMS Collaboration], Phys. Rev. Lett. **99**, 091301 (2007).
[21] E. Behnke *et al.* [COUPP Collaboration], Science **319**, 933 (2008).
[22] G. J. Alner *et al.* [UK Dark Matter Collaboration]Phys. Lett. B **616**, 17 (2005).

Directional Neutron Detection and TPC Development at LLNL

M. Heffner

7000 East Ave, L-050, Livermore, CA 94551

mheffner@llnl.gov

Abstract. LLNL is involved with a number of TPC projects spanning basic science to homeland security. This talk outlines the TPC work at LLNL and specifically focuses on the neutron TPC.

1. Introduction

A number of TPC projects are now underway at Lawrence Livermore National Laboratory (LLNL) and there is currently a ramp up in the infrastructure both in equipment and people to support these efforts. In place are high pressure vessels for xenon studies up to 50bar, larger vessels up to 100 litters at 10bar, clean room facilities, extensive electronics development, dedicated lab space and a assortment of radioactive sources.

1.1. Xenon TPC

Xenon is an attractive gas for both neutrino-less double beta decay and dark matter searches because of the low average ionisation energy. In the case of neutrino-less double beta decay this provides better energy resolution, and if the statistical limit could be reached it would only be a few times worse than germanium, but much cheaper and therefore scaleable to much larger detectors. The initial goal of this work at LLNL would be to look at negative ion drift as a method to count charges and try to achieve the statistical limit while maintaining tracking [1].

1.2. Fission TPC

The development of advanced fission reactors needs better nuclear data to simulate accurately the effects of engineering choices. Cross sections on the major actinides have been identified as a major source of error [2] for the simulations and this project is aimed at reducing the systematic error on neutron induced fission cross sections. The tradition method of using a fission chamber has perhaps reached a limit with respect to reducing systematic errors, and the construction of a TPC to address the systematic errors has now been funded. LLNL is the lead institution for the construction of this TPC that will be run at the LANSCE facility at Los Alamos National Laboratory.

2. Neutron TPC

A detector that can identify neutrons and measure the direction that the neutrons arrived from could be useful for homeland security purposes. The goal of the neutron TPC[3] project is to build such a detector from a hydrogen filled TPC.

Journal of Physics: Conference Series **179** (2009) 012010 doi:10.1088/1742-6596/179/1/012010

The detection principle is very simple. A fast neutron (~2MeV) from a fission source elastically scatters with the hydrogen in the TPC to produce a proton track in the gas. This proton is tracked and the direction of the proton is correlated with the direction of the incoming neutron. Although the kinematics of the scatter blurs the correlation, it only takes about 10 neutron scatters to reduce the cone of uncertainty to the neutron source down to about 16 degrees. The He(n,p)T reaction has also been considered because of the favorable kinematics, but the cross section is about 4.5 less which increases dwell time and cost.

Perhaps one of the most significant tasks in this project is not the physics of the detection process, but the design of a TPC that will be used in the field by non-scientists. The TPC will have to withstand vibration, jolts, temperature swings, possible mis-configuration, abuse and automatic analysis. There are also addition constraints such as weight, size, power consumption, and safety issues in transporting and using the TPC. Each of these items appear to be solvable, but considerable effort will have to go into them and may have lessons that are useful for the design of other TPCs.

A prototype have been built and operated in the lab at LLNL. It has an active volume of about 36 liters and is filled with hydrogen at 2bar absolute. A 83μCi ^{252}Cf source was used as a neutron source as its spectrum is similar to that of other spontaneous fission sources of interest. This source was moved to different locations within and outside the lab to access the performance of the TPC. At 10meters the source could be located with in an hour with the small prototype, and with dwell times of a few hours the source could be located at 20 meters and behind concrete walls. A larger TPC could reduce the dwell times significantly.

3. Electronics

A significant portion of the cost to design and build a TPC is in the electronics. At LLNL we are building a set of electronics with all off-the-shelf components. This is possible because of advances in what is available. For example high speed ADCs with high-density packing, very high performance FPGA and very small discrete components (0201 package). The advantage is that the cost per channel is low even in small quantity. This is good for small or prototype detectors, and it is highly configurable due to the FPGA. The analog portion has been designed and built, and is now in testing. The digital part has been designed and is not in the layout phase. The goal is to have a working set by the end of the year.

4. Conclusion

The TPC effort at LLNL is growing and has something to offer back to the TPC community in electronics development and the accompanying software.

References

[1] W. Chinowsky et al. Ionization imaging: A new method to search for 0-nu decay. Nucl. Instrum. Meth., A580:829–835, 2007

[2] G. Aliberti, G. Palmiotti, M. Salvatores, et al., Annals of Nuclear Energy 33, (2006)

[3] I. Jovanovic, M. Heffner, L. Rosenberg, N. S. Bowden, A. Bernstein, D. Carter, M. Foxe, M. Hotz, M. Howe, A. Myers, and C. Winant, Directional detection of special nuclear materials using a neutron time projection chamber, Accepted by IEEE Transactions on Nuclear Science (2008).

Micromegas μTPC for direct Dark Matter search with MIMAC

F. Mayet[1], O. Guillaudin[1], C. Grignon[1], C. Koumeir[1], D. Santos[1], P. Colas[2], I. Giomataris[2]

[1] LPSC, Université Joseph Fourier Grenoble 1, CNRS/IN2P3, Institut Polytechnique de Grenoble, 53 avenue des Martyrs, 38026 Grenoble, France

[2] IRFU/DSM/CEA, CE Saclay, 91191 Gif-sur-Yvette cedex, France

E-mail: `Frederic.Mayet@lpsc.in2p3.fr`

Abstract. The MIMAC project is a multi-chamber detector for Dark Matter search, aiming at measuring both track and ionization with a matrix of micromegas μTPC filled with ^{3}He and CF_4. Recent experimental results on the first measurements of the Helium quenching factor at low energy (1 keV recoil) are presented, together with the first simulation of the track reconstruction. Recontruction of track of α from Radon impurities is shown as a first proof of concept.

1. Introduction

Tremendous experimental efforts on a host of techniques have been made in the field of direct search of non-baryonic dark matter. However, going further required an unambigious positive signal in order to distinguish a geniune WIMP event from backgrounds (mainly neutrons and γ-rays). The most promising strategy is to search for a correlation of the signal with the motion of the detector with respect to the galactic Dark Matter halo. This can either be an annual modulation, due to the motion of the earth around the sun or a strong direction dependence of the incoming WIMP, toward the Cygnus constellation to which points the sun's velocity vector [1, 2]. Several projects aiming at directional detection of Dark Matter are being developped [3, 4, 5, 6].

Gaseous μTPC detectors present the privileged features of being able to reconstruct the track of the recoil following the interaction, thus allowing to access both the energy and the track properties (lenght and direction). A precise measurement of the energy of the recoil is the starting point of any background discrimination.

2. The MIMAC project

The MIMAC project is a multi-chamber detector for Dark Matter search. The idea is to measure both track and ionization with a matrix of micromegas μTPC filled with ^{3}He and CF_4. The use of these two gases is motivated by their privileged features for dark matter search. In particular, a detector made of such targets will be sensitive to the spin-dependent interaction, leading to a natural complementarity with existing detectors mainly sensitive to scalar interaction, in various SUSY models, e.g. non-universal SUSY [7, 8]. Moreover, as shown in [8], a 10 kg ^{3}He dark matter

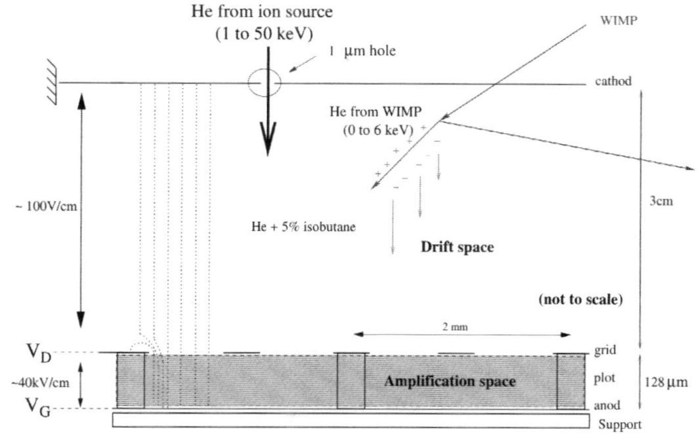

Figure 1. Sketch of the Micromegas μTPC used for the Quenching factor measurement (not to scale).

detector[1] with a 1 keV threshold (MIMAC) would present a sensitivity to SUSY models allowed by present cosmology and accelerator constraints. This study highlights the complementarity of this experiment with most of current spin-dependent experiments : proton based detectors as well as ν telescopes.
Using both ^{3}He and CF_4 in a patchy matrix of μTPC opens the possibility to compare rates for two atomic masses, and to study neutralino interaction separately with neutron and proton as the main contribution to the spin content of these nuclei. With low mass targets, the challenge is also to measure low energy recoils, below 6 keV for Helium, by means of ionization measurements. Low pressure operation of the MIMAC detector will open the possibility to discriminate neutralino signal and background on the basis of track features and directionality.

3. Ionization quenching factor measurement

The ionization quenching factor (IQF) is defined as the fraction of energy released through ionization by a recoil in a medium compared with its kinetic energy. Measuring IQF, especially at low energies, is a key point for Dark Matter detectors, since it is needed to evaluate the nucleus recoil energy and hence the WIMP kinematics.
The energy released by a particle in a medium produces in an interrelated way three different processes: i) ionization, producing a number of electron - ion pairs, ii) scintillation, producing a number of photons through de-excitation of quasi-molecular states and iii) heat produced essentially by the motion of nuclei and electrons. The fraction of energy given to electrons has been estimated theoretically [9] and parametrized by [10].
In the last decades an important effort has been made to measure the IQF in different materials : gases [11], solids [12, 13] and liquids [14], using different techniques. The use of a monoenergetic neutron beam has been explored in solids with success [15, 16]. However in the low energy range the measurements are rare or absent for many targets (e.g. Helium) due to ionization threshold of detectors and experimental constraints.
We have developped an experimental setup devoted to the measurement of low energy (a few keV) ionization quenching factor. The purpose is twofold : measuring for the first time the Helium quenching factor and performing very low energy measurements to test the prototype cell (resolution, threshold, ...) in the range of interest for Dark Matter.

[1] or the equivalent mass of CF_4

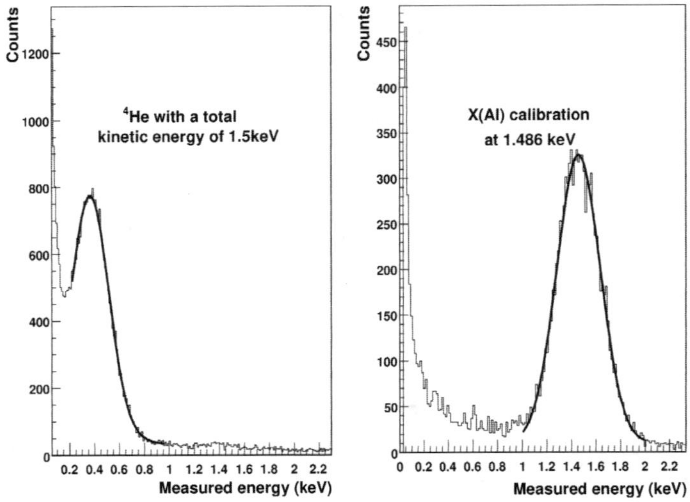

Figure 2. Spectra of 1.5 keV kinetic energy ^{4}He (left) and 1.486 keV X-ray of ^{27}Al (right) in ^{4}He +5% C_4H_{10} mixture at 700 mbar.

The experimental set-up is the following : an Electron Cyclotron Resonance Ion Source [17] with an extraction potential from a fraction of kV up to 50 kV, is coupled to Micromegas (micromesh gaseous) detector via a 1 μm hole with a differential pumping (fig. 1).
Ion energy values have been previously checked by a time-of-flight measurements through a 50 nm thick N_4Si_3 foil [18]. The ionization produced in the gas has then been measured with a Micromegas (micromesh gaseous) detector [19] adapted to a cathode integrated mechanically to the interface of the ion source. The μ-TPC is composed of a bulk type Micromegas [20], in which the grid and the anode are built and integrated with a fixed gap, 128 μm for measurements between 350 and 1300 mbar. The electric fields for the drift and the avalanche have been selected to optimize the transparence of the grid and the gain for each ion energy. Typical applied field were 100 V/cm for the drift and a voltage of 450 V for the avalanche. The drift distance between the cathode and the grid was 3 cm, large enough to include the tracks of ^{4}He nuclei of energies up to 50 keV. These tracks, of the order of 6 mm for 50 keV, are roughly of the same length than the electrons tracks produced by the X-rays emitted by th ^{55}Fe source used for calibration.

In order to measure the quenching factor of ^{4}He in a gas mixture of 95% of ^{4}He and 5% of isobutane (C_4H_{10}) we proceed as follows: i) the ionization given by the Micromegas was calibrated by the two X-rays (1.486 and 5.97 keV) at each working point of the Micromegas defined by the drift voltage (V_d), the gain voltage (V_g) and the pressure, ii) the number of ions per seconde sent was kept lower than 25, to prevent any problem of recombination in primary charge collection or space charge effect.

Two different calibration sources have been used : the 1.486 keV X-rays of ^{27}Al produced by alpha particles emitted by a source of ^{244}Cm under a thin foil of aluminium and a standard ^{55}Fe X-ray source giving the 5.9 keV K_α and the 6.4 keV K_β lines. These two lines, as they were not resolved by our detector, have been considered as a single one of 5.97 keV, taking into account their relative intensities. The IQF of a recoil is then the ratio between the ionization energy and the recoil kinetic energy. In such a way, the IQF compares the nuclei ionization efficiency with respect to the electrons. The ionization spectra of 1.5 keV ^{4}He nuclei and of electrons of

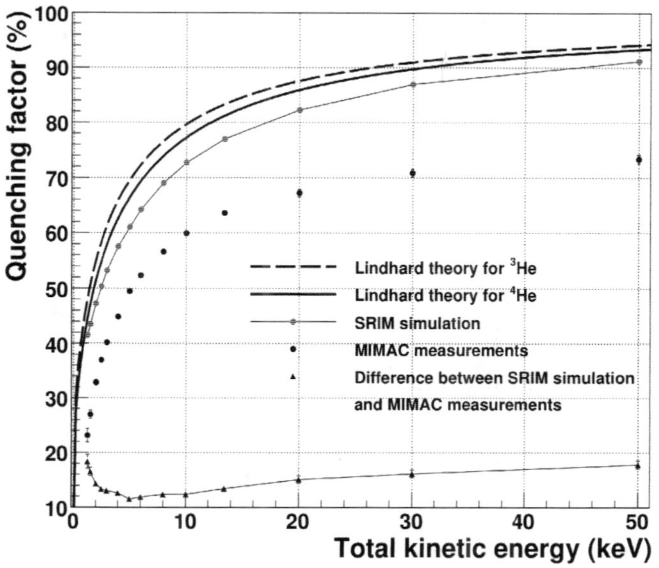

Figure 3. Helium ionization quenching factor as a function of ^{4}He kinetic energy (keV). Lindhard theory prediction (^{4}He in pure ^{4}He and ^{3}He in pure ^{3}He), parametrized as in [10], is presented and compared with SRIM simulation results (solid line and points) in the case of ^{4}He in ^{4}He +5% C_4H_{10} mixture. Measured quenching factor are presented at 700 mbar with error bars included mainly dominated by systematic errors. The differences between data and SRIM simulation are shown by triangles.

roughly the same energy (1.486 keV) are shown on Figure 3.

The measurement reported [21] have been focused on the low energy ^{4}He IQF. Figure 3 presents the results at 700 mbar compared with the Lindhard theory for ^{4}He ions in pure ^{4}He and with respect to the SRIM simulation [22] for ^{4}He in the same gas mixture used during the measurements. We observe a difference between the SRIM simulation and the experimental points of up to 20% of the kinetic energy of the nuclei, shown in fig. 3. This difference may be assigned to the scintillation produced by the ^{4}He nuclei in ^{4}He gas. This difference is reduced at lower pressures due to the fact that the amount of scintillation is reduced when the nuclei mean distance is increased giving a lower production probability of eximer states. More details may be found in [21].

As shown in [23] energy resolution of Micromegas μTPC has been measured down to 1 keV. Moreover, it does not depend on pressure and it does not affect the number of expected events for Dark Matter, even at low energy. Threshold as low as 300 eV (ionization) has been reached. As presented in [21], measurements have been done in various experimental conditions, showing a clear increase at lower gas pressure and also at lower isobutane percentage (the quencher). Hence, the ionization signal is expected to increase with Helium purity, which is needed for Dark Matter search, and also at the lowest pressures, which is compulsory for directional detection.

4. Track simulation and measurement

The second step of a dark matter project aiming at directional detection is to show the possibility to reconstruct a 3D track. This is a key point as the required exposure is decreased by an order of magnitude between 2D read-out and 3D read-out [2].

The 3D reconstruction strategy chosen for the MIMAC project is the following (see fig. 4) : i)

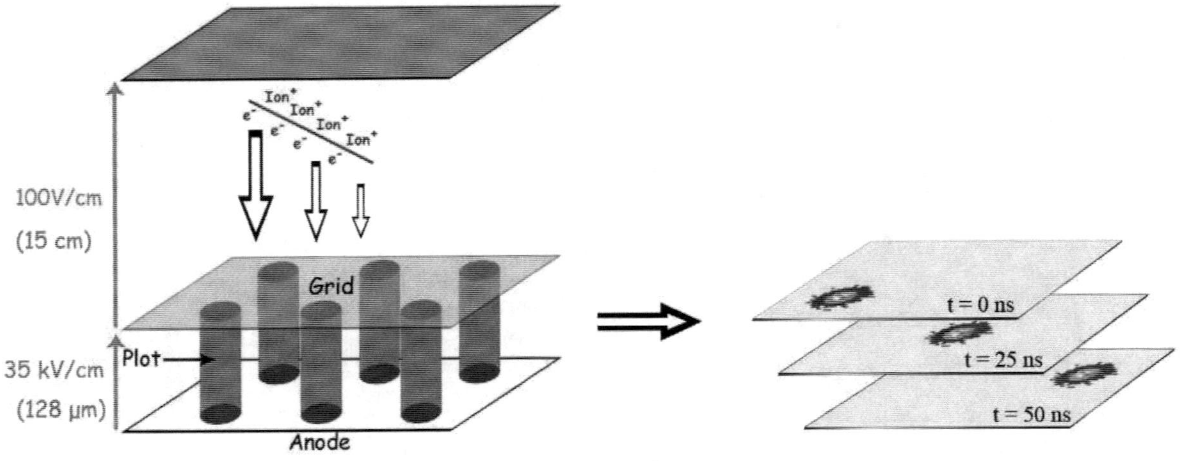

Figure 4. Track reconstruction in MIMAC. The anode is scanned every 25 ns and the 3D track is recontructed, knowing the drift velocity, from the serie of images of the anode.

the electrons from the track are projected on the anode thus allowing to access information on x and y coordinates, ii) the anode is read every 25 ns, iii) knowing the drift velocity[2], the serie of images of the anode allows to reconstruct the 3D track.
The anode, composed of x and y stripes with 200 μm pitch, is read every 25 ns by a dedicated ASIC ship designed and built at LPSC.
Simulation of the principle of the 3D track reconstruction can be achieved with a rather good resolution, of the 0.3mm on the track length and below 4° on θ and ϕ [24].
Very first track reconstruction have been recently obtained with 5 MeV α from Radon impurities. Figure 5 presents an α 3D track obtained at 800 mbar in a Helium + 5 % isobutan mixture. This can be taken as a proof of the principle of track reconstruction strategy chosen for this project. Same results should be achieved in the near future with low energy recoils (below 6 keV), in the range of interest for Dark Matter.

5. Conclusion

In summary, this first measurement of Helium ionization in Helium down to energies of 1 keV recoil opens the possibility to develop a Helium μTPC for Dark Matter. 3D track reconstruction proof on principle, from simulation and α data is a first step toward a gaseous TPC aiming at directional detection of Dark Matter.

6. References

[1] Spergel D N, Phys. Rev. D **37** (1988) 1353
[2] Green A M & Morgan B, Astropart. Phys. **27**, 142 (2007)
[3] Alner G J *et al.*, Nucl. Instr. Meth. A **555** (2005) 173
[4] Sciolla G *et al.*, arXiv:0806.2673
[5] Santos D *et al.*, arXiv:astro-ph/0701230
[6] Collar J I & Giomataris Y, Nucl. Instrum. Meth. A **471** (2000) 254
[7] Mayet F *et al.*, Phys. Lett. **B538** (2002) 257
[8] Moulin E *et al.*, Phys. Lett. **B614** (2005) 143
[9] Lindhard J *et al.*, Mat. Fys. Medd. K. Dan. Vidensk. Selsk. 33 (1963) 1-42.
[10] Lewin J D & Smith P F, Astropart. Physics **6** (1996) 87-112

[2] The value 26 μm/ns is used for simulation at 1 bar[24]

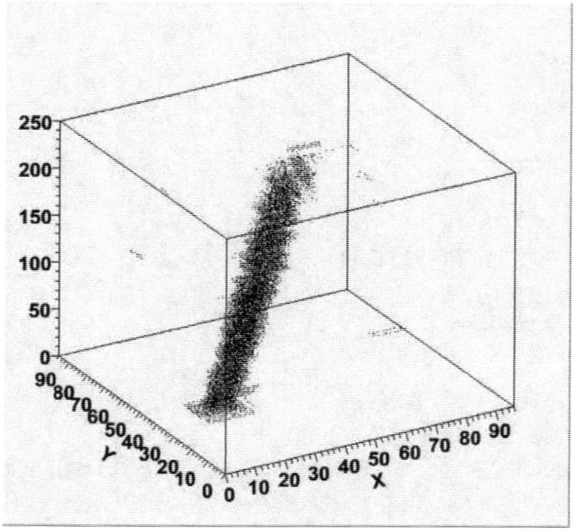

Figure 5. First result on 3D track reconstruction obtained with 5.59 MeV α from Radon impurities, at 800 mbar in a Helium + 5 % isobutan mixture.

[11] Verbinski V V & Giovannini R, Nucl. Instr. and Meth. **114** (1974) 205
[12] Gerbier G *et al.*, Phys. Rev. **D42** (1990) 3211
[13] Jones K W & Kraner H W, Phys. Rev. **A11** (1975) 1347
[14] Aprile E *et al.*, Phys. Rev. Lett. **97** (2006) 081302
[15] Jageman T *et al.*, Nucl. Instr. and Meth. **A551** (2005) 245
[16] Simon E *et al.*, Nucl. Instr. and Meth. **A507** (2003) 643
[17] Geller R, Electron Cyclotron Resonance Ion Sources and ECR plasmas, Bristol and Philadelphia : Institute of Physics Publishing, 1996
[18] Mayet F *et al.*, to appear
[19] Giomataris Y *et al.*, Nucl. Instr. and Meth. **A376** (1996) 29
[20] Giomataris Y *et al.*, Nucl. Instr. and Meth. **A560** (2006) 405
[21] Santos D, Mayet F, Guillaudin O *et al.*, arXiv:0810.1137
[22] Ziegler J *et al.*, http://www.srim.org
[23] Guillaudin O, Mayet F, Grignon C *et al.*, Proc. of the 4th symposium on large TPCs for low energy rare event detection, Paris, Dec. 2008, Journal of Physics: Conference Series
[24] Grignon C *et al.*, in preparation

Journal of Physics: Conference Series **179** (2009) 012012 doi:10.1088/1742-6596/179/1/012012

Low energy measurements with Helium Micromegas μTPC

O. Guillaudin[1], F. Mayet[1], C. Grignon[1], C. Koumeir[1], D. Santos[1], P. Colas[2], I. Giomataris[2]

[1] LPSC, Université Joseph Fourier Grenoble 1, CNRS/IN2P3, Institut Polytechnique de Grenoble, 53 avenue des Martyrs, 38026 Grenoble, France
[2] IRFU/DSM/CEA, CE Saclay, 91191 Gif-sur-Yvette cedex, France

E-mail: olivier.guillaudin@lpsc.in2p3.fr

Abstract. The measurement of the ionization produced by particles in a medium presents a great interest in several fields from metrology to particule physics and cosmology. The ionization quenching factor is defined as the fraction of energy released by ionisation by a recoil in a medium compared with its kinetic energy. At low energy, in the range of a few keV, the ionization falls rapidly and systematic measurement are needed. We have developped an experimental setup devoted to the measurement of low energy (keV) ionization quenching factor for the MIMAC project. The ionization produced in the gas has been measured with a Micromegas detector filled with Helium gas mixture.

1. Introduction

A nuclear recoil moving at very low energies can be used to detect rare events such as neutrino coherent interactions or non-baryonic dark matter signatures. The direct detection of these non-baryonic particules is based on the detection of nuclear recoils coming from elastic colisions on different targets. Ionization is one of the most important channel to detect such nuclear recoils. The ionization quenching factor is defined as the fraction of the kinetic energy released through ionization by a recoil in a medium. In the last decades, an important effort has been made to measure the IQF in different materials : gases [1], solids [2, 3] and liquids [4], using different techniques. However, in the low energy range, the measurements are rare or absent for many targets due to ionization threshold of detectors and experiment contraints.
We have developped an experimental setup devoted to the measurement of low energy (a few keV) ionization quenching factor for the MIMAC project [5, 6].

The measurements reported in this work have been done with a micromegas detector filled with a gas mixture of 95% of ^{4}He and 5% of isobutane (C_4H_{10}) at different pressures. The Micromegas detector was calibrated using X-rays produced by fluorescence on different targets. To produce nucleus moving with a controlled energy in the detection volume, we have developped calibrated Electron Cyclotron Resonance Ion Source (ECRIS) [7] which produced ions with energy from a fraction of one keV up to 50 keV. The ECRIS source was coupled to a Micromegas detector [8].

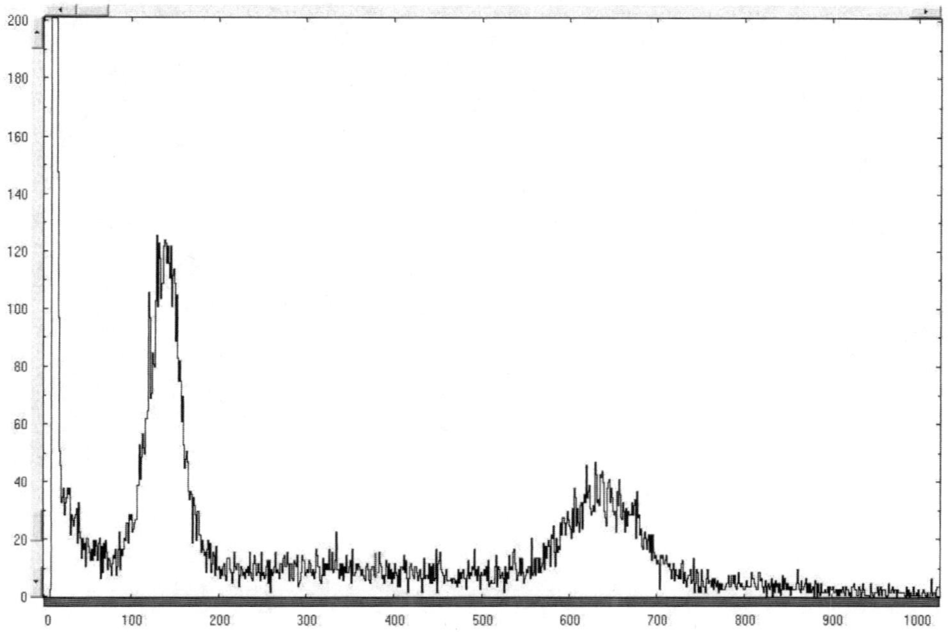

Figure 1. Response of ^4He + 5% C_4H_{10} Micromegas to Aluminium (1.486 keV) and Iron (6.4 keV) fluorescence X-rays at 1000 mbar, with a resolution (FWHM) of 29,4 % and 15,3 % respectively.

2. Experimental setup and detector calibration

The micromegas used in this work was a standard Bulk Micromegas [9], in which the mesh and the anode are built and integrated with a fixed 128 μm gap. The gap between the copper anode and stainless steel mesh is maintained by cylindrical pilars every 3 mm. This gap is well adapted for working gas pressure between 350 and 1300 mbar. The drift distance between the cathode and the mesh was 3 cm, large enough to include the tracks of nuclei of energies up to 50 keV at 350 mbar. The active area was 100×100 mm^2.

We used an individual power supply for each electrode (mesh and anode) with cathode drift to the ground, allowing to independently vary the different fields. The NIM High Voltage module (ISEG 223M) presents a high stability and high precision of the output voltage setting and measurement (about 10 mV). The setting parameters, the voltage and the current controle are performed by remote control via a LabView interface. Typical applied field were 100 V/cm for the drift and a voltage of 450 V for the avalanche. With these values, an optimum collection efficiency is achieved and the electron drift velocity is comparatively low.

The electronic system is composed of a dedicated charge pre-amplifier and of a commercial spectrocopy amplifier. Due to the low drift velocity of electron in such a low drift field (100 V/cm), the integration time constant of this pre-amplifier should be large enough to assure a total colection charge and an output proportionnal to the total energy of the event. The calibration constant for the preamplifier is measured by applying test pulse on a test capacity. A PCI-bus multichannel analyzers with internal 8k ADC is used to convert incoming signals.

For each measurement, the gas vessel is previously pumped using a turbomolecular pump and a dry primary vacuum pump. Then the detector is filled with gas mixture at the working pressure and operates with a circulating and controlled flux. To correct this low gain variation, a calibration is performed just before each energy measurement with the ion source.

Figure 2. Electron Cyclotron Resonance Ion Source (ECRIS) and the 45 degree spectrometer (yellow part) coupled to the Micromegas detector.

Due to the high transparency of ^{4}He gas for X-rays energy above 2 keV, the detector shows a very low efficiency to X-ray from a ^{55}Fe source used for calibration purpose, especially at low pressure. So, to avoid using high intensity radioactive source and to calibrate in a realistic time, we have developped a high intensity multi energies X-ray system based on the production of fluoresence photons on different high purity targets. The system used a pulsed miniature X-rays generator inside the gas vessel with maximum flux equivalent to a 2 mCi radioactive source and the following targets with the corresponding X-ray energies : Aluminium (1.486 keV), Titanium (4.504 keV), Iron (6.4 keV), Copper (8.1 keV). Figure 1 shows a typical spectrum for Aluminium (1.486 keV) and Iron (6.4 keV) targets with a resolution (FWHM) of 29,4 % and 15,3 % respectively. The energy threshold is close to 300 eV.

3. The ion source

The ECRI ion source is a plasma device designed and built by the LPSC SSI team to provide charged ions at low velocities from a fraction of a keV to 50 keV. One of the important features of this facility is the availability of very low and stable currents of a few picoamp. In the present configuration, the beam is adapted to the 45 degree spectrometer by an Einzel lens which has been tuned with the faraday cup. A micrometer aperture (1 μm in diameter) selects a part of this beam which is injected in the Micromegas detector and produced about 25 ions per second. According to the gas injected is the plasma chamber, the source is able to produce many type of beams : proton, ^{3}He, ^{4}He, ^{19}F.
To calibrate the ion source, a 50 nm thick Silicon Nitride membrane (Si_3N_4) was used as interface between the ion source and the gas chamber. A time of flight measurement has been performed under vacuum, using two channeltrons. One of them detecting the low energy electrons extracted from the Si_3N_4 menbrane by the ions and the other one detecting the ions for 6 different known positions. This setup allows to measure the energies of the ions just after the foil. With this configuration a first energy measurement with micromegas was performed. In such a way, we could verify that the energy measured by TOF were the same as those indicated by the extraction potentiel values in kV for the 1+ charge state ions with the 1 μm diameter aperture interface.

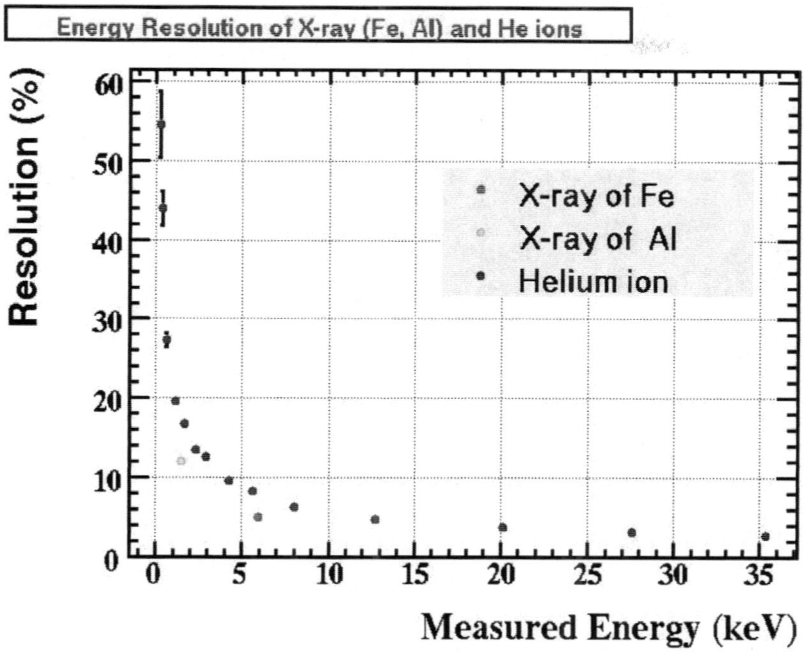

Figure 3. Energy resolution for X-rays and ^4He ions in a ^4He + 5% C_4H_{10} Micromegas detector at 700 mbar.

4. Low energy ion measurements

Using the photoelectron produced by fluorescence X-rays as calibration, we can measure ionization energies produced by the recoils. The ionization quenching factor of a recoil will be the ratio between this energy and its kinetic energy. The measurement reported [10] have been focused on the low energy ^4He IQF [8].
At the same time, it was possible to extract the Micromegas energy resolution for a wide range of incoming ion energies. Ionization energies have measured for ions with kinetic energy down to 1 keV. Figure 3 presents the evolution of the resolution (FWHM) at 700 mbar for measured ionization energies of ^4He ions from 300 eV to 35 keV. We can note that the resolution is slightly better for X-rays. At low energy, down to 1 keV, the energy resolution is about 50%, but this does not affect the number of expected events for Dark Matter search [8].
Using the possibility to produce recoils of different states of charge (0,+1,+2 for ^4He) with the same kinetic energy, it is also possible to confirm that the ionization energy does not depend of the state of charge of the recoil. This was clearly demonstrated with ^4He recoil with 50 keV of kinetic energy. Measurement with neutral recoils have been obtained with the Si_3N_4 window because ions are neutralized when passing through the membrane. Moreover, this experimental setup allow to measure the quenching factor for other different recoils wich could be used for dark matter search such as : ^3He in ^3He or ^{19}F in CF_4.
Concerning ^3He, which is a rather expensive gas, a clean close loop gas system is under construction to allow gas recovery and storage after the measurements.

5. Conclusion

In summary, we have demonstrated the possibility to measure ionization energy for ^4He recoils down to 1 keV with a compatible energy resolution for Dark Matter search. For the next step, this measurement will be extended to low gas pressure and to different recoil such as ^3He of ^{19}F.

Journal of Physics: Conference Series **179** (2009) 012012 doi:10.1088/1742-6596/179/1/012012

This measurement is particularly important to better understand the ionization response of gas detectors.

6. References

[1] Verbinski V V & Giovannini R, Nucl. Instr. and Meth. **114** (1974) 205
[2] Jones K W & Kraner H W, Phys. Rev. **A11** (1975) 1347
[3] Gerbier G *et al.*, Phys. Rev. **D42** (1990) 3211
[4] Aprile E *et al.*, Phys. Rev. Lett. **97** (2006) 081302
[5] Mayet F *et al.*, Phys. Lett. **B538** (2002) 257
[6] Moulin E *et al.*, Phys. Lett. **B614** (2005) 143
[7] Geller R, Electron Cyclotron Resonance Ion Sources and ECR plasmas, Bristol and Philadelphia : Institute of Physics Publishing, 1996
[8] Mayet F, Guillaudin O, Grignon C *et al.*, Proc. of the 4th symposium on large TPCs for low energy rare event detection, Paris, Dec. 2008, Journal of Physics: Conference Series
[9] Giomataris Y *et al.*, Nucl. Instr. and Meth. **A560** (2006) 405
[10] Santos D, Mayet F, Guillaudin O *et al.*, arXiv:0810.1137

Journal of Physics: Conference Series **179** (2009) 012013 doi:10.1088/1742-6596/179/1/012013

The Gotthard and MUNU TPCs: from the past to the future?

Jean-Luc Vuilleumier
LHEP, University of Bern, Sidlerstrasse 5, 3012 Bern, Switzerland
E-mail: jean-luc.vuilleumier@lhep.unibe.ch

Abstract. The performance of two similar gas TPCs used in past low energy particle physics experiments are briefly described. The first was installed in the Gotthard underground laboratory to search for the neutrinoless double beta decay of ^{136}Xe. The second, MUNU, was operated near the Bugey nuclear reactor to study $\overline{\nu}_e e^- \rightarrow \overline{\nu}_e e^-$ scattering. The advantages of the design are presented, along with the limitations. Ideas to improve the performance, for future experiments in low energy, low rate, particle physics are presented. The emphasis is put on the search for the neutrinoless double beta decay of ^{136}Xe.

1. Introduction

Gas TPCs have turned out to be very useful tools in low energy, rare event, particle physics. Among these are the Gotthard TPC, used for the search of the neutrinoless double beta decay of ^{136}Xe, and MUNU, operated near the Bugey nuclear reactor to study $\overline{\nu}_e e^- \rightarrow \overline{\nu}_e e^-$ scattering. From the experience gained in the operation of these detectors, modifications are proposed in the following. They should lead to enhanced performances of TPCs in future experiments, achieving a better energy resolution and a better background suppression thanks to a better event signature. The emphasis is put on the search for neutrinoless double beta decay in ^{136}Xe. In that case good energy resolution is essential to distinguish neutrinoless double beta decay from the allowed two neutrino one. These remarks are part of the thinking going on within the EXO collaboration [1] towards a search of this decay using a gas TPC, as an alternative to the baseline option with a liquid xenon TPC. They complement the presentation of D. Sinclair in these proceedings, where ion tagging is discussed. Similar ideas could however also apply to a next generation gas TPC for the study of low energy solar neutrinos in a TPC filled with CF_4 [2].

1.1. The Gotthard TPC

The Gotthard TPC [3] was installed in the Gotthard underground laboratory. It was contained in a copper vessel surrounded by a lead shielding (fig. 1). It was filled with xenon gas enriched to 68 % in ^{136}Xe and 5 % CH_4 as a quencher. The TPC was operated at a pressure of 5 bar and had a fiducial volume of 180 l (5.3 kg of xenon). The fiducial volume was delimited by a cathode at the bottom, and a grid at the top. The charge was amplified behind the grid on an anode plane made from 20 μm thick wires with a 4.95 mm spacing, separated by potential wires. The anode signal was used to measure the total charge deposited in the fiducial volume, from which the energy is reconstructed. A pick-up plane having perpendicular x and y strip with

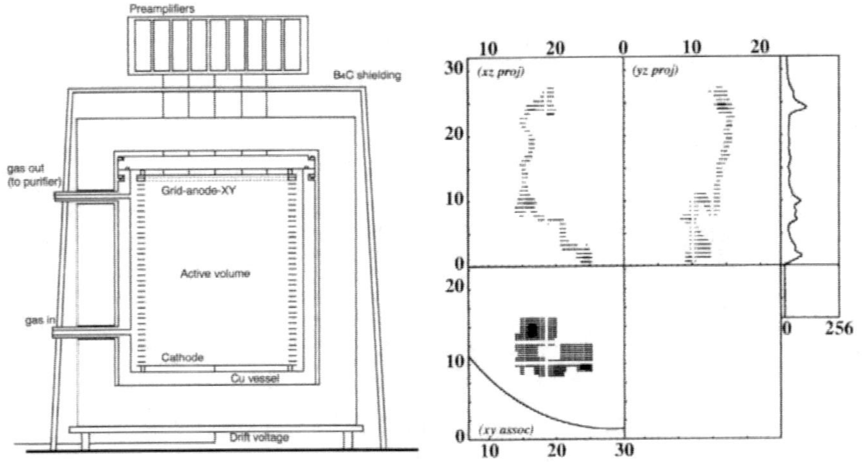

Figure 1. Left: the Gotthard TPC inside its lead shielding; right: x-z and y-z projections, the anode signal, as well as the reconstructed x-y projection, of a double beta decay candidate event.

3.95 mm pitch, placed behind the grid, measured the induced signal. With that information, combined with the z coordinate reconstructed from the drift time, the x-z and y-z projections of events were obtained. A good double beta decay candidate is a single continuous track within a radial fiducial volume with enhanced energy deposition at both ends. An example is shown in fig. 1. The energy calibration and the resolution were measured using radioactive sources. The double escape peak of the 2614 keV line in ^{208}Tl was particularly useful. Extrapolating with the square root of the energy, an energy resolution of $\sigma(E)/E = 2.7$ % was determined at 2.48 MeV, the energy Q_0 released in the decay. This is an average over the drift depth of the TPC, since no absolute z coordinate was available. No correction for charge losses along the drift path could be applied. Using ^{210}Po α decays on the cathode, with fixed z, an energy resolution of $\sigma(E)/E = 1.1$ % at 2.48 MeV was reconstructed, indicative of what can be achieved.

The experiment found no excess of events above background around Q_0, within energy resolution, and a limit of $T^{0\nu}_{1/2} > 4.4 \times 10^{23}$ yr was derived for neutrinoless double beta decay.

1.2. The MUNU TPC

The MUNU TPC [4] was inspired from the Gotthard one. It retained the same structure with, however, a larger fiducial volume (1000 l, fig. 2). It was filled with CF_4 at 3 bar for most measurements. CF_4 has a relatively low Z, meaning reduced multiple scattering. The TPC was contained in an acrylic vessel, immersed in a tank filled with liquid scintillator and viewed by photomultipliers. The scintillator served as a cosmics veto and as an anti-Compton detector. It was very useful in further constraining the background. An other advantage of MUNU over the Gotthard TPC was the read-out electronics, which brought a better spatial resolution.

Single electron candidate events were single continuous tracks within a fiducial radius, with increased energy deposition at one end. An example is shown in fig. 2. An excess of electrons from the reactor was found, in agreement within uncertainties with predictions of the standard model. An upper limit of $\mu_\nu < 9 \times 10^{-11}$ μ_B on the magnetic moment of the neutrino was derived.

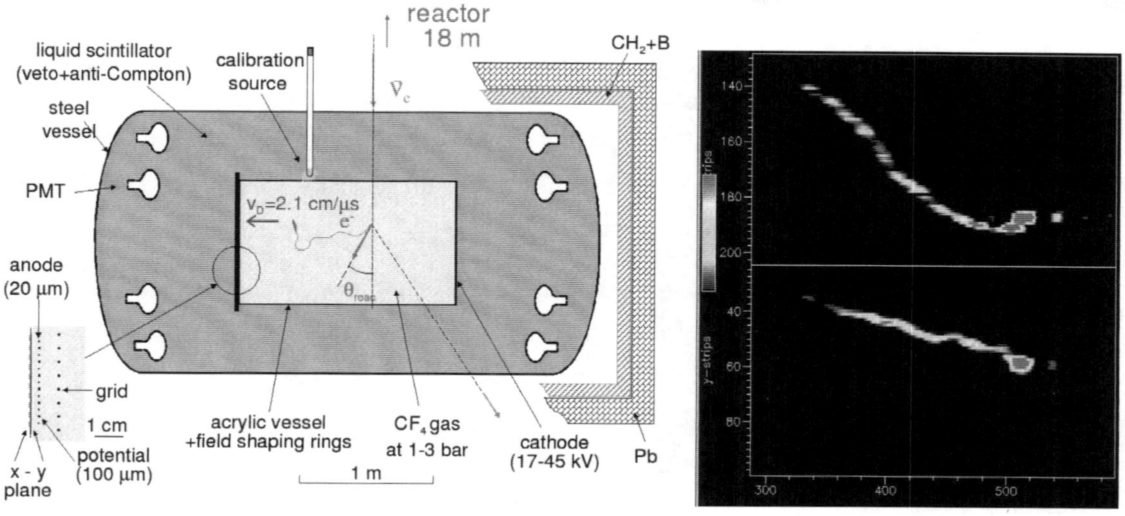

Figure 2. Left: The MUNU TPC; right: the x-z and y-z projections of a single electron track.

2. Possible improvements

To go to larger, more sensitive gas TPCs, improvements are required. Here are some ideas:

- A better reliability is required. The anodes wires were a problem. Here it is proposed to replace the anode plane and the grid by a Micromegas pattern [5].
- An absolute z-determination is necessary. This can be achieved by measuring the primary scintillation light, providing an absolute reference for the drift time and thus the z coordinate. This would help in correcting for the charge losses along the drift, and lead to an improved energy resolution. Moreover it would allow to define a fiducial cut in z, not only a radial one as in the Gotthard TPC and in MUNU. It would make possible a 3-D fiducial cut, eliminating entirely α and β backgrounds from the walls, which are generally found to be significantly dirtier than the detector medium.
- A better spatial resolution is necessary. In the Gotthard TPC, and even in MUNU, the spatial resolution was not limited by diffusion, lateral or longitudinal, the ultimate limitation, but by electronic noise problems and by the granularity. Segmented Micromegas [6] could bring a substantial improvement. Improved spatial resolution would allow a better event selection, primarily by a better identification of the increased ionization at the end of an electron track.

3. Micromegas with xenon

In Neuchâtel and in Bern we have tested Micromegas read-out patterns with xenon gas using a small TPC [7]. The drift volume has a diameter of 10 cm and a length of 20 cm. The compact Micromegas consists of an anode and a grid, compacted "two in one" and separated with cylindrical pin spacers. The anode is a plain copper plated board with 9 cm of diameter. Kapton spacers with 250 μm diameter by 100 μm height or more and placed every 1 mm, are laid on it by conventional lithography. The height defines the amplification gap size. The grid is made of stainless steel wires, with 20 μm of diameter, woven with a pitch of 53 μm. It was tensioned delicately and glued with a thin glue layer to the outer part of the anode. The grid is pressed against the spacers.

No amplification sufficient to do tracking was obtained in pure xenon. It was found however that an addition of CF_4, even minute, gives a significant improvement. CF_4 was chosen as an additive because it does not absorb light. Measured gains and FWHM energy resolution at 5.5 keV versus grid voltage with a 100 μm gap at atmospheric pressure are shown in fig. 3. The Xe(98 %)CF_4(2 %) mixture appears particularly attractive.

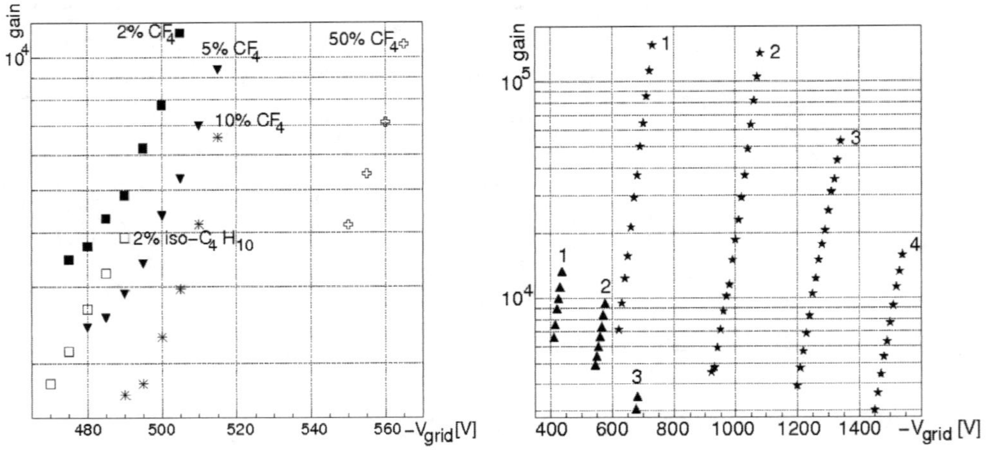

Figure 3. Left: gain obtained with a Micromegas with 100 μm gap at atmospheric pressure for various gas mixtures in function of the grid voltage; right: gain at various pressures (bar) for 75 μm (▲) and 225 μm (★) gaps in Ar(90 %)CH_4(10 %).

Also it was shown that it is possible to increase the pressure to several bars if the Micromegas gap is increased as well. Gains in Ar(90 %)CH_4(10 %) are shown in fig. 3. The Micromegas with 225 μm gap was also successfully operated in Xe(98 %)CF_4(2 %) at pressures up to 4 bar. Gains of the order of 10^3 were comfortably achieved at 4 bar. Clearly operation at 5-10 bar is possible.

4. Segmented Micromegas

Using photolithographic techniques it is possible to produce segmented Micromegas with a great flexibility in the geometry [6]. The anode is subdivided in pixels, which are read out individually or, connected together in various patterns, in groups. The grid can be segmented as well, in a coarser manner. The grid signal provides additional spatial information allowing to remove ambiguities. Fine granularity can be envisioned, with correspondingly good spatial resolution.

5. The primary scintillation light

Xenon, argon and CF_4 are known to be good scintillators. Xenon has a broad emission spectrum centered around 175 nm [8]. Several wavelength shifters can bring it to visible blue light. TPB, for instance, has been reported to have a fluorescence yield as large as 40 % [9]. TPB can be doped to polystyrene which can be evaporated to form thin films on light guides. In these impinging UV light will be shifted to blue light, a large part of which being trapped by total reflection, and reaching the end with good efficiency. There, it can be detected, for instance with photomultipliers.

6. A large gas TPC concept

All this leads to a large gas TPC for the search of double beta decay in ^{136}Xe having the general structure depicted in figure 4. The central TPC has a Micromegas read-out system to measure the ionization charge with optimal spatial resolution. It is filled with a mixture of Xe(98 %)CF_4(2 %) at 5-10 bar. This gas mixture should be transparent to xenon's UV primary scintillation light. The primary xenon scintillation light is shifted to the blue in the TPB doped light guides and detected in photomultipliers. The central TPC vessel is immersed in a liquid scintillator anti-Compton detector, which is also efficient at detecting the annihilation gamma rays from positrons. This allows to suppress the background from e^+e^- pairs, which exhibit the same topology as double beta events.

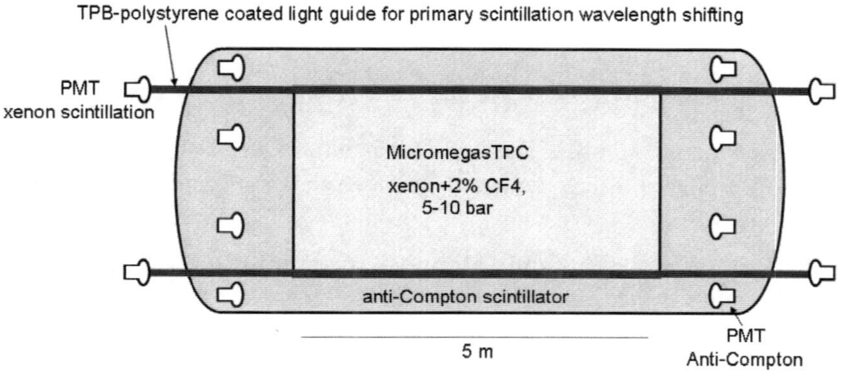

Figure 4. A possible layout for a gas TPC to search for double beta decay in ^{136}Xe, with an anti-Compton and a read-out system for the primary xenon scintillation light.

All this, along with an installment deep underground and a careful selection of clean materials for the construction of the detector, should ensure a very high background suppression. With the dimensions given in the figure, masses of order 1 t can be achieved. However, R&D is necessary on some points to entirely demonstrate the feasibility and optimize the design.

Acknowledgments

Many thanks to Razvan Gornea and Martin Auger, with whom many of the ideas in this presentation were discussed.

[1] Leonard D *et al.* (EXO collaboration) 2008 *Nucl. Inst. Meth.* A **591** 4909
[2] Daraktchieva Z *et al.* 2008 *J. Phys. G: Nucl. Part. Phys.* **35** 125107
[3] Luescher R *et al.* 1998 *Phys. Lett.* B **434** 407
[4] Daraktchieva *et al.* 2005 *Phys. Lett.* B **615** 153
[5] Giomataris Y *et al.* 1996 *Nucl. Inst. Meth.* A **376** 29
[6] Oliveira R *CERN* private communication
[7] Ounalli L *et al.* 2009 *JINST* **4** P01001
[8] Jortner J *et al.* 1965 *J. Chem. Phys.* **42** 4250
[9] Kinsey D N *et al.* 2004 *Nucl. Inst. Meth.* A **516** 475

Optical readout of secondary scintillation from liquid argon generated by a thick gas electron multiplier

P.K.Lightfoot [a*], G.J. Barker [b], K. Mavrokoridis [a], Y.A. Ramachers [b], N.J.C. Spooner [a]

[a] Department of Physics and Astronomy, University of Sheffield, Hicks Building, Hounsfield Road, Sheffield, S3 7RH, UK
[b] Department of Physics, University of Warwick, Coventry, CV4 7AL, UK

Abstract

For the first time secondary scintillation, generated within the holes of a thick gas electron multiplier (TGEM) immersed in liquid argon, has been observed and measured using a silicon photomultiplier device (SiPM).

250 electron-ion pairs, generated in liquid argon via the interaction of a 5.9KeV Fe-55 gamma source, were drifted under the influence of a 2.5KV/cm field towards a 1.5mm thickness TGEM, the local field sufficiently high to generate secondary scintillation light within the liquid as the charge traversed the central region of the TGEM hole. The resulting VUV light was incident on an immersed SiPM device coated in the waveshifter tetraphenyl butadiene (TPB), the emission spectrum peaked at 460nm in the high quantum efficiency region of the device.

For a SiPM over-voltage of 1V, a TGEM voltage of 9.91KV, and a drift field of 2.5KV/cm, a total of 62 ± 20 photoelectrons were produced at the SiPM device per Fe-55 event, corresponding to an estimated gain of 150 ± 66 photoelectrons per drifted electron.

Key words: Electroluminescence, secondary scintillation, proportional light, argon, gas electron multiplier, silicon photomultiplier, neutrino, dark matter.

Corresponding author: Dr. Phil Lightfoot, Department of Physics and Astronomy, University of Sheffield, Hicks Building, Hounsfield Road, Sheffield S3 7RH, UK,

Tel: +44(0)1142224533, Fax: +44(0)1142223555, E-mail:p.k.lightfoot@sheffield.ac.uk

1. Introduction

The liquid argon time projection chamber (TPC) is a leading technology candidate for the type of large scale underground detector widely acknowledged to be the next-generation project in the areas of neutrino oscillation physics, astro-particle physics and proton decay. The current challenge is to establish methods of tracking and homogeneous calorimetry combining the excellent performance of a liquid argon TPC with a practical implementation that is cost effective and readily scalable up to the benchmark of 100kton.

Charge losses over long drift distances in liquid argon mean that amplification of charge is required. Gas electron multipliers (GEMs) [1], TGEMs [2], Micromegas [3], and Bulk Micromegas [4], have evolved over the last 20 years and are now routinely used within large volume targets. Significant charge amplification within the liquid has however proven difficult to achieve in a practical design [5] leading to the consideration of two-phase liquid argon volumes where the charge amplification using TGEM's occurs in the gas phase above the liquid volume. These designs bring their own problems however and there is a risk the performance could be compromised on the largest scales by the requirements of precise liquid levelling, extreme cooling stability and space-charge effects at the liquid-gas interface.

There is motivation therefore to investigate alternative ways in which a single phase liquid argon volume could be used for 3D tracking and in particular the possibility of using scintillation light readout rather than charge. The successful operation of single phase liquid argon TPC's based on light readout via, for example the imaging of TGEM planes with SiPM arrays, would significantly reduce the complexity of design and allow modular construction of large arrays with readout in any plane.

This report describes the first measurement of secondary scintillation light produced within the holes of a TGEM and viewed by a SiPM device all within a single phase liquid argon target, and builds on an earlier study in which SiPM performance was assessed [6] at cryogenic temperatures.

2. Experimental Details

The liquid argon cryostat used in all tests was contained within a pressurised liquid nitrogen volume. The internal assembly, shown in figure 1, consisted of a 1mm^2 SiPM device positioned directly above the centre of a 65mm diameter TGEM, located above a 20mm drift region defined by a woven steel cathode at the base of the assembly.

The TGEM was manufactured from a double faced copper clad FR-4 epoxy resin glass reinforced composite plate of thickness 1.5mm, hole diameter 1mm, pitch 1.5mm, only the central region of the TGEM being perforated. Two 5mm diameter polystyrene cylinders were fitted between the SiPM and the TGEM to isolate the SiPM device from background scintillation light from the argon target as shown in figure 1.

Although the principle aim was the measurement of secondary scintillation from liquid argon using the SiPM device, satisfactory performance of the TGEM was first confirmed by reading out the charge produced in the gas phase of a double phase argon system at cryogenic temperature. In all cases the bottom face of the TGEM was grounded, a drift field between 1 and 4KV/cm created by applying negative potential to the steel cathode. The amplification field within the TGEM was controlled at the top TGEM electrode and the charge signal, read from the top TGEM electrode, passed through an Amptek A250 charge sensitive preamplifier to an Acqiris CC108 PCI acquisition system triggered through a discriminator unit.

The optical signal was readout using a SensL series 1000 SiPM connected to an external preamplifier. Device characteristics are listed in table 1. The output signal from the SiPM

Journal of Physics: Conference Series **179** (2009) 012014 doi:10.1088/1742-6596/179/1/012014

preamplifier was split, one signal passing directly to an input channel on the Acqiris unit. The other signal was connected through an Ortec 572 shaping amplifier with a 10μs integration time, to a N417 discriminator and then to the trigger input of the Acqiris unit.

Table 1. Characteristics of SensL series 1000 SiPM device and associated pre-amp.

Pixel size	Number of cells	Geometric efficiency %	Maximum gain	Pixel recovery time (ns)	Spectral range (nm)	$V_{breakdown}$ at 25°C
20μm	848	43	8×10^{6}	40ns	400 - 700	28.2V

Prior to all tests the inner chamber was evacuated to 1×10^{-8} mbar and baked to 60°C. For all low temperature measurements as liquid nitrogen was slowly added to the cryogenic jacket, the pressure of argon gas within the target was maintained at 1 bar until liquefaction eventually occurred. At this point the cryogenic jacket was filled with liquid nitrogen and pressurized to between 3 and 3.5 bar thereby increasing the boiling point of liquid nitrogen to the point at which liquid argon could condense within the target at 1 bar.

Argon emits VUV scintillation light at 128nm. Light collection was increased by coating the SiPM device with a waveshifter to shift direct 128nm VUV light to 460nm visible light [7] and into the sensitive high quantum efficiency region of the device [6].

Fe-55 (5.9KeV gamma emitter) was positioned at the centre of the steel woven cathode within the drift region, the source collimated using a thin steel sheet to ensure gamma interactions were limited to the vertical axis through the SiPM device. The average number of primary electrons-ion pairs generated from a 5.9 KeV Fe-55 source assuming a W-value of 23.6eV [8] was 250. The ratio of the charge collected at the top TGEM electrode or the number of secondary photons detected by the SiPM to the charge generated within the drift region was then used to determine the charge gain and optical gain respectively of the TGEM when operated in charge amplification mode or scintillation mode.

Figure 1. Internal assembly used to detect secondary scintillation in liquid argon.

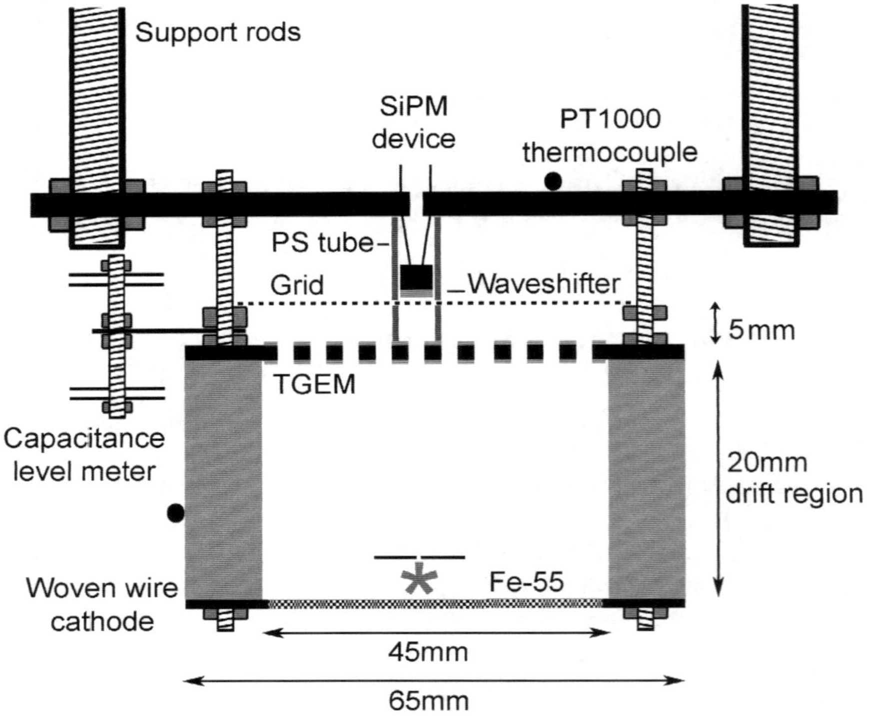

3. Results

In order to fully characterise the assembly, it was essential to measure the performance of the SiPM device, to ensure that the TGEM was operating correctly and that the argon was of sufficiently high purity to allow charge drift. Following this, the field within the TGEM was varied and secondary scintillation measured as a function of the TGEM field.

3.1 Limit of SiPM device linearity due to saturation effects at high LED photon flux rates

Due to the digital nature of operation, at any time a proportion of the total number of pixels of the SiPM device will be in a state of recovery. This characteristic places a limit on the extent of linearity of the device especially for high intensity pulses of similar time scale to the recovery time. For electron recoils within liquid argon, approximately 23% of all photons are produced within the first 6ns following an interaction, the remaining 77% forming an exponential distribution with a 1590ns time constant [9]. An evaluation of the linearity of the output signal from the SiPM with increasing light intensity was performed at -196°C. Full experimental details are contained in reference [6]. A pulsing circuit [10] was connected to an externally mounted light emitting diode (LED) producing 5ns rise-time 15ns decay-time light pulses of variable intensity, which were passed through a fibre optic cable to the SiPM device held on a support structure within the target. For a fixed over-voltage, the intensity of the 460nm LED pulse was gradually increased and the output signal recorded.

Measurements show linearity of the output signal is maintained at low incident photon fluxes irrespective of the over-voltage. Data shown in figure 2 indicates a 5% deviation from linearity to occur for a 1V over-voltage for incident fluxes containing in excess of 1500 photons. In all future measurements using argon each scintillation pulse was assessed to ensure that the photon flux within any 40ns window did not exceed the stated 5% deviation limit for the selected over-voltage value.

Figure 2. Linearity of a SiPM device at high LED photon flux rate.

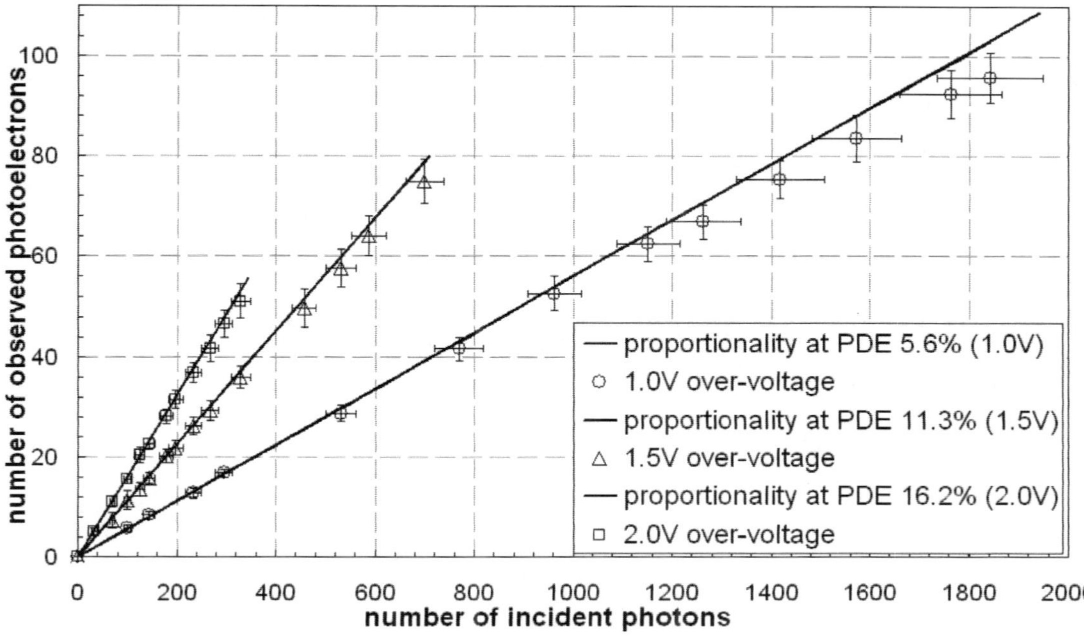

Journal of Physics: Conference Series **179** (2009) 012014 doi:10.1088/1742-6596/179/1/012014

3.2 Cryogenic operation of the TGEM in charge amplification mode in the vapour phase of a double phase argon system

Liquid argon was condensed within the chamber and the TGEM operated in charge gain mode in order to confirm that both the magnitude of the drift field and the degree of purification were sufficient to enable charge transfer through the liquid. By careful addition of argon the liquid level was set so as to produce corresponding fields within the vapour phase below the TGEM of 4.0KV/cm, and within the liquid of 2.5KV/cm.

Measurements, shown in figure 3, were made of the charge gain as a function of the amplification field within the TGEM in 1 bar cold saturated argon gas above the liquid phase. For amplification fields in excess of 18KV/cm, a photopeak was identified due to the internal 5.9KeV Fe-55 source, shown in figure 4. For a constant amplification field, these events were stopped by reduction of the drift field as shown in figure 5, thereby demonstrating that they originated from within the liquid.

Figure 3. Charge gain versus potential across TGEM in 1 bar saturated argon gas above the liquid phase of a double phase system for a 2.5KV/cm constant drift field within the liquid. Room temperature charge gain in 3 bar argon is shown for comparison.

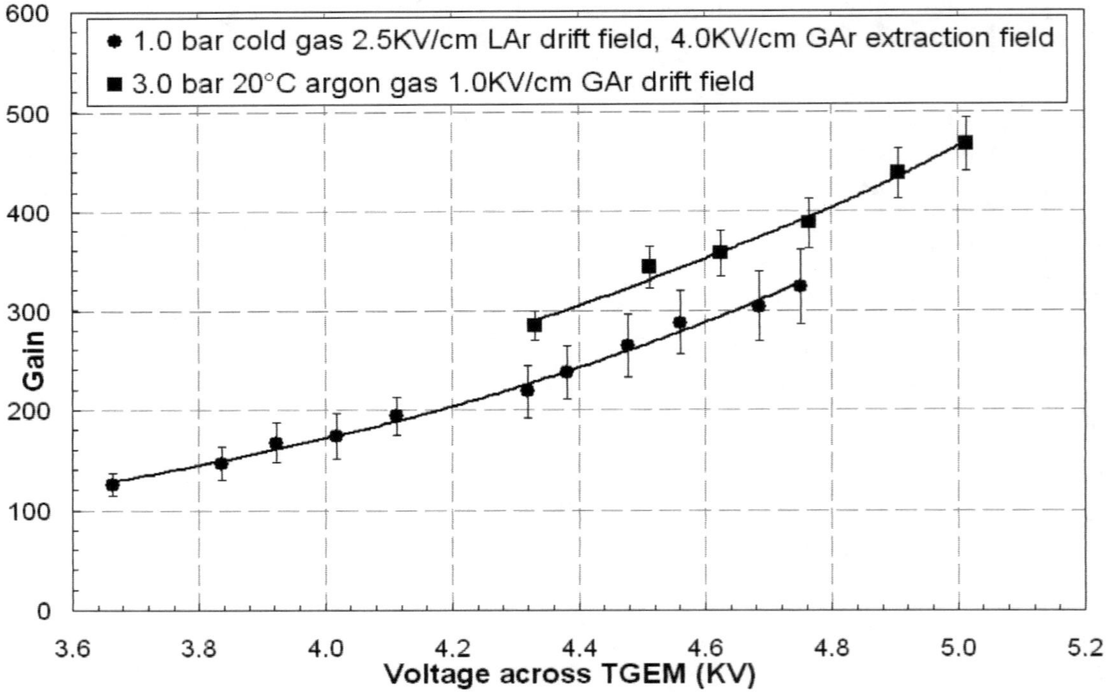

Figure 4. Charge spectrum from Fe-55 in the cold 1 bar gaseous argon phase of a double phase system. (V_{TGEM} = 4.685KV, Drift field in liquid = 2.5KV/cm and in gas = 4.0KV/cm, gain = 303, 171388 total events.)

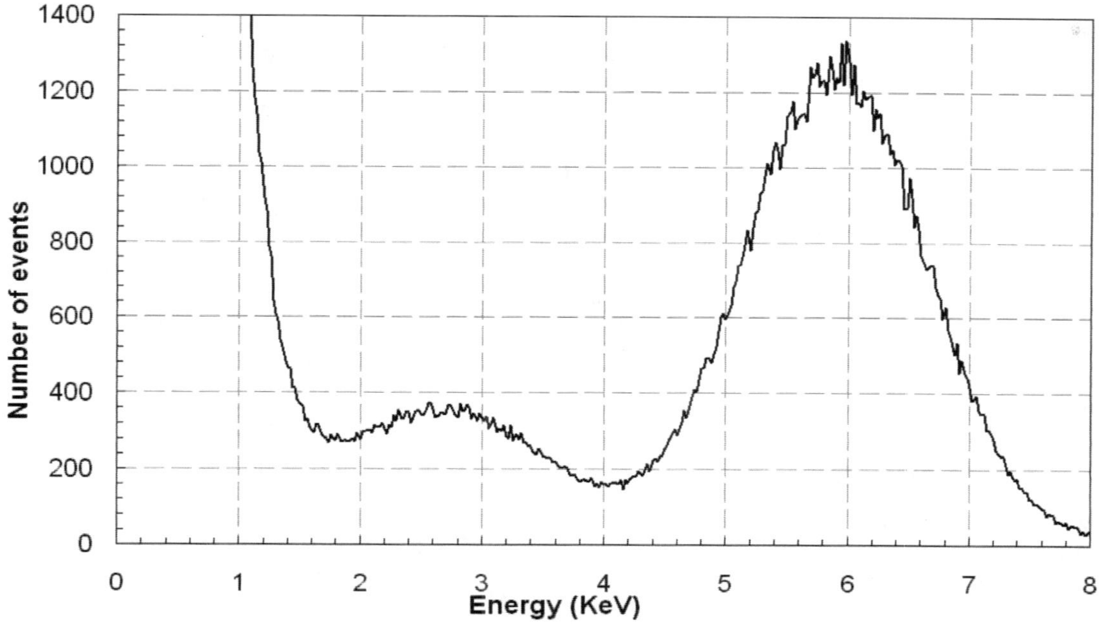

Figure 5. Charge gain versus drift field within liquid argon for a fixed TGEM potential of 4.75KV.

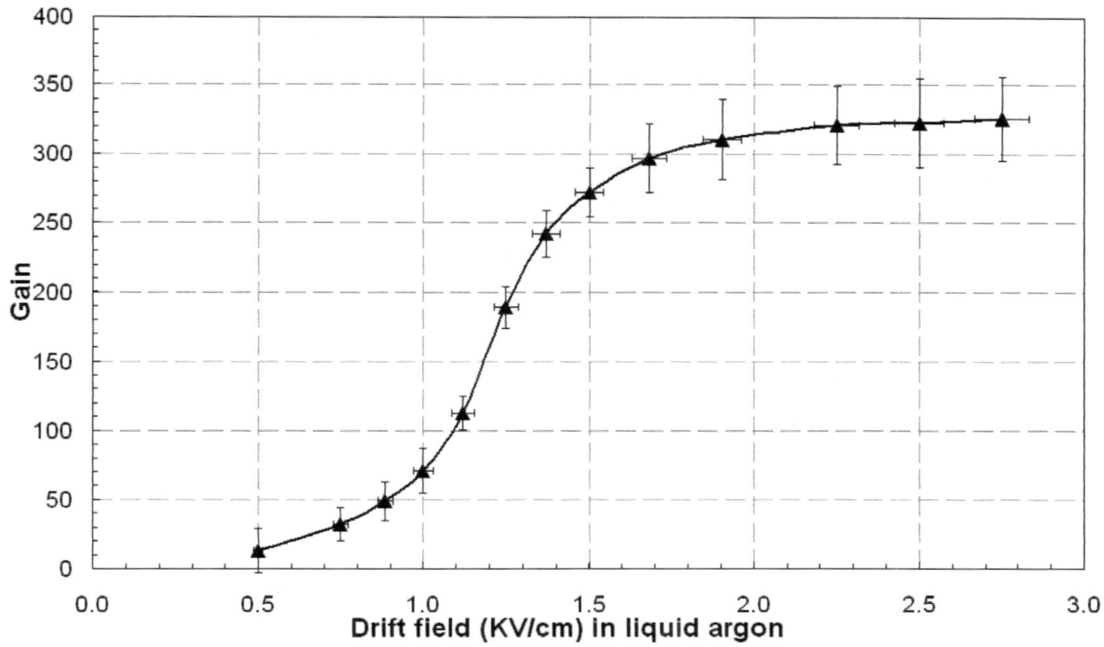

3.3 Cryogenic measurement of secondary scintillation generated within a TGEM in the vapour phase of a double phase argon system using a SiPM device

The SiPM device, now with a breakdown voltage of 24.2V at -189°C [6] was operated at a bias voltage of 25.2V corresponding to an over-voltage of 1V, in 1 bar saturated argon vapour. For a constant drift field of 2.5KV/cm within the liquid phase, the electric field between the TGEM electrodes was steadily increased and the photoelectron pulse distribution produced by the SiPM within a 10μs data acquisition time window recorded.

As the field within the TGEM holes increased, secondary scintillation was observed between 2.1KV and 3.4KV. Figure 6 shows a secondary scintillation spectrum for Fe-55 for 2.16KV across the TGEM, the presence of both the argon escape peak (electron binding energy of 3.2KeV) and approximately 12 photoelectron peaks above the discriminator level providing calibration for the 5.9KeV Fe-55 peak. Figure 7 shows light collection as a function of TGEM voltage. Although a linear fit could be applied to the distribution, figure 3 suggests an exponential relationship to be more appropriate. Although the origin of the scintillation pulses can be ascertained from the presence of both the escape peak and the Fe-55 full absorption peak, further evidence can be gleaned from the scintillation response as the drift field is reduced as shown in figure 8.

Figure 6. Secondary scintillation spectrum from Fe-55 in the cold 1 bar gaseous argon phase of a double phase system. (SiPM over-voltage=1V, V_{TGEM} = 2.16KV, Drift field in liquid = 2.5KV/cm and in gas = 4.0KV/cm, 5.9KeV peak corresponding to 86 photoelectrons.)

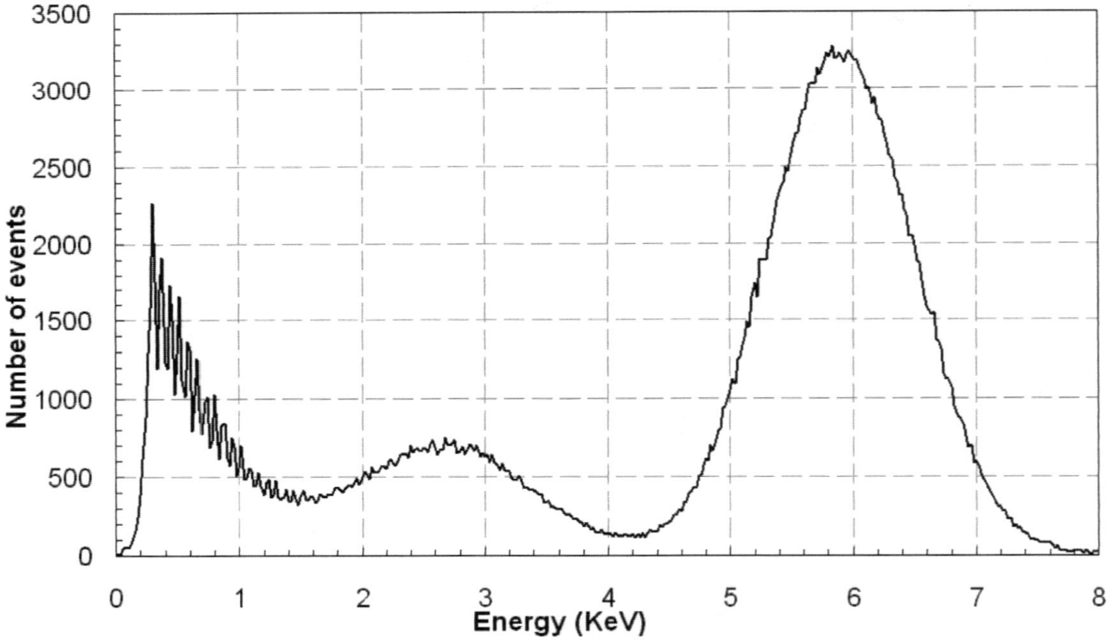

Figure 7. Secondary scintillation in 1 bar -189°C gaseous argon as a function of the voltage across a TGEM due to the passage of charge generated via the interaction of an Fe-55 source in the liquid phase within a 2.5KV/cm drift field.

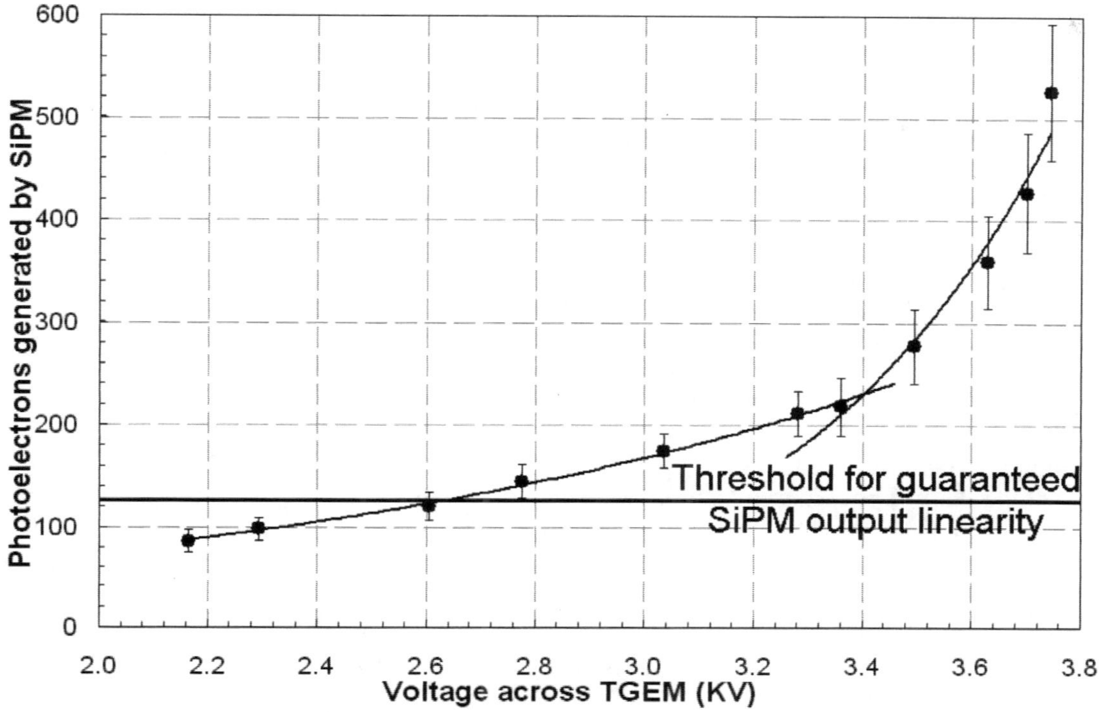

Figure 8. Secondary scintillation versus drift field within liquid argon for a fixed TGEM potential of 3.36KV operating in the cold gas phase of a double phase target.

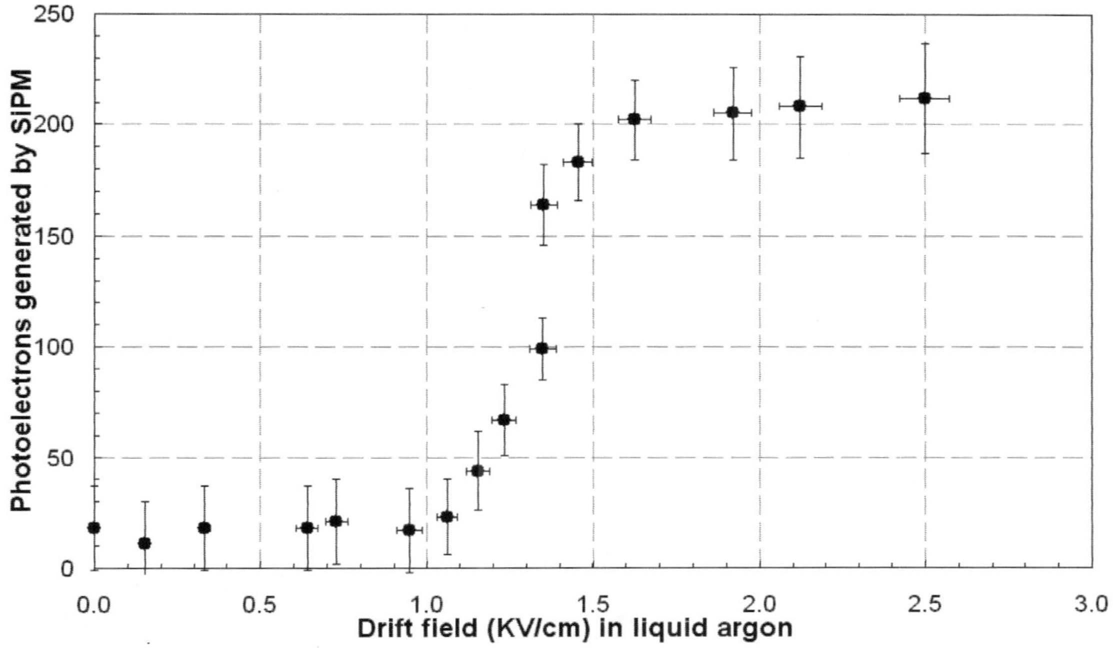

Journal of Physics: Conference Series **179** (2009) 012014 doi:10.1088/1742-6596/179/1/012014

3.4 Secondary scintillation generated within a TGEM measured using a SiPM device with both immersed in a single phase liquid argon system

Sections 3.2 and 3.3 successfully demonstrated both cryogenic operation of the SiPM device and successful charge transfer through purified liquid argon. Whilst charge amplification in cold argon vapour has already been successfully achieved [11], exhaustive testing failed to generate any measurable charge gain from a TGEM submerged in liquid argon. Following directly from double phase testing, liquefaction was continued until both the SiPM and TGEM were submerged. The drift field was set to 2.5KV/cm and the voltage across the TGEM was slowly increased. At voltages across the TGEM in excess of 8KV the rate and magnitude of the photoelectron pulses became more pronounced. By careful adjustment of both the discriminator threshold level, and the use of a low pass hardware filter, a peak believed to be due to the Fe-55 source was observed at 9.85KV. As the voltage was increased the resolution decreased until at 10.15KV the peak became almost impossible to separate from the background contribution.

The resolution of the Fe-55 signal is degraded in liquid compared to previous measurements in gas. This is believed to be as a consequence of the long 10μs integration time required to fully acquire the slow component time constant of 1590ns [9]. Because of the extended decay time compared with for example xenon, each triggered event contains, in addition to an Fe-55 pulse, contributions from primary scintillation due to radioactive background interaction, DCR effects, and crucially photons created due to sparking within the TGEM operating close to the breakdown threshold. With increasing TGEM field, the contribution from these sources increased, reducing the resolution of the Fe-55 pulse above 9.91KV and effectively masking it above 10.15KV. Figure 9 shows a characteristic pulse attributed to Fe-55 at 9.91KV between the TGEM electrodes, whilst figures 10 shows spectra taken for TGEM voltages of 9.91KV. The number of photoelectrons generated at the SiPM as a function of the voltage across the TGEM is shown in figure 11.

Figure 9. A secondary scintillation light pulse produced by an Fe-55 source in a liquid argon system. (SiPM over-voltage=1V, V_{TGEM} = 9.91KV, Drift field in liquid = 2.5KV/cm.)

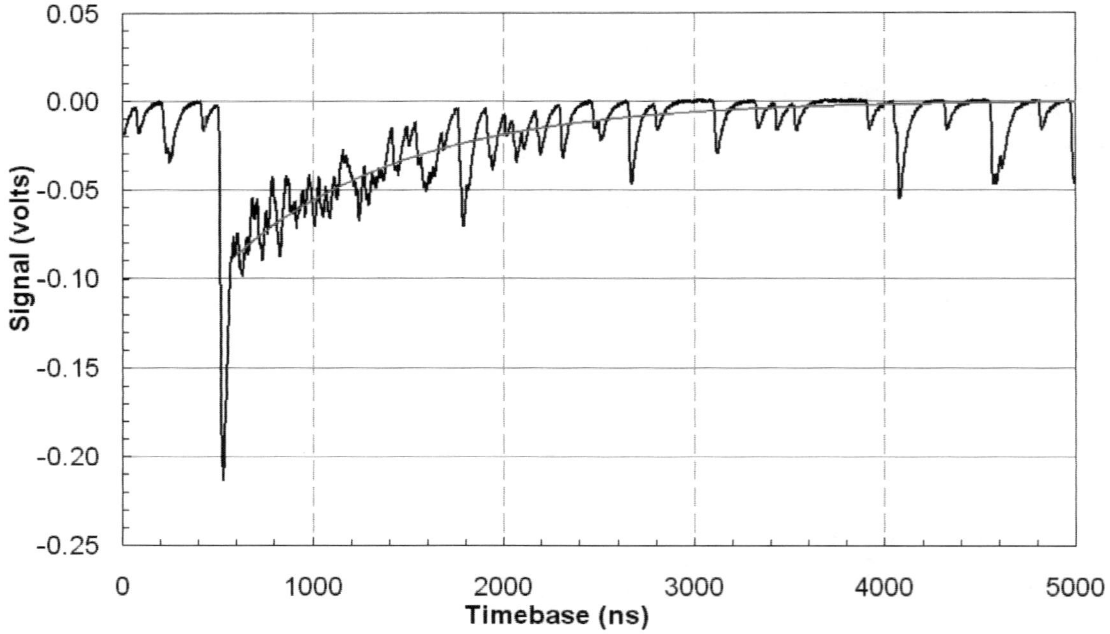

Figure 10. Secondary scintillation spectrum from Fe-55 in a liquid argon system. (SiPM over-voltage=1V, V_{TGEM} = 9.91KV, Drift field in liquid = 2.5KV/cm, 5.9KeV peak corresponding to 62 photoelectrons, 1037508 total events.)

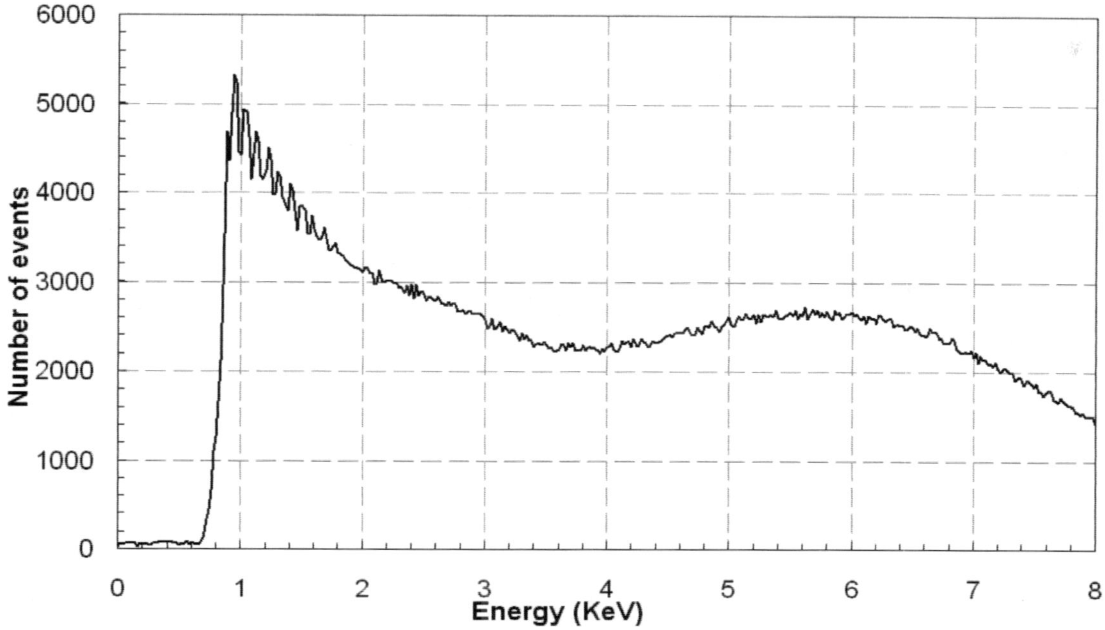

Figure 11. Secondary scintillation generated in liquid argon within a TGEM viewed by a submerged SiPM as a function of voltage across a TGEM at a constant 2.5KV/cm drift field.

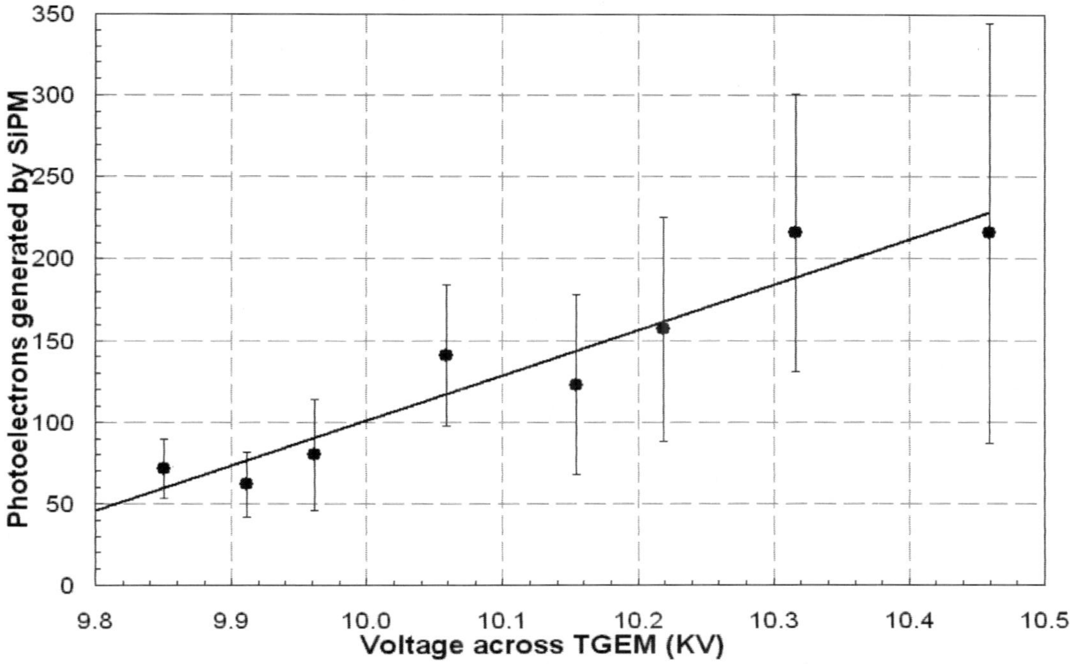

The drift field was then reduced from 2.5KV/cm, and the number of photoelectrons collected at the SiPM corresponding to the Fe-55 peak recorded for two applied voltages across the TGEM. Although a reduction in photoelectron collection is observed for both cases, the effect was more pronounced for 9.91KV due to a combination of reduced background noise and greater energy resolution of the Fe-55 peak position. Results are shown in figure 12.

Figure 12. Secondary scintillation corresponding to the Fe-55 peak versus drift field within liquid argon for fixed TGEM potentials of both 9.91KV (red) and 10.15KV (black).

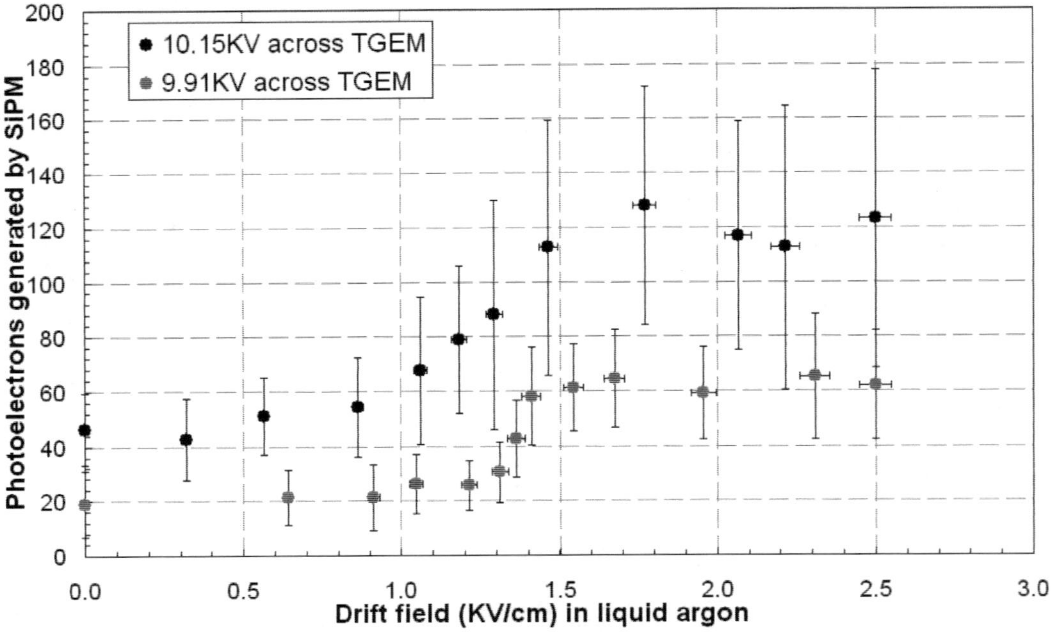

Finally an estimate was made of the photon yield produced at the TGEM shown in figure 13.

Figure 13. Estimated secondary scintillation gain in liquid argon as a function of voltage across the TGEM.

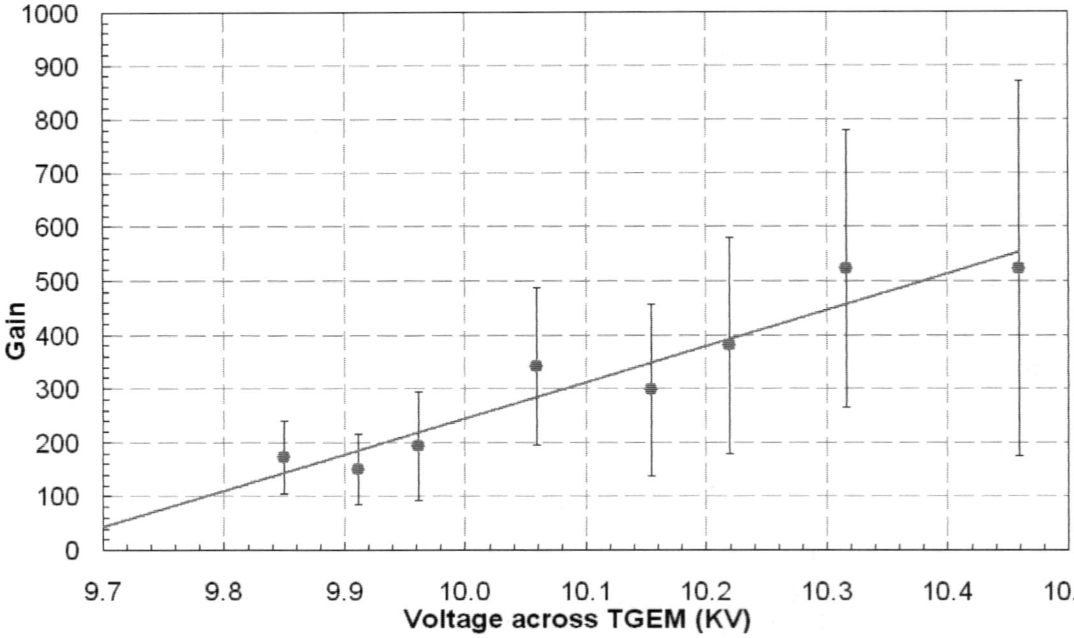

4. Discussion

Studies of charge amplification [5][12] in pure liquid argon have failed to yield any reliable gain. Irrespective of the choice of charge readout device, the maximum charge gain drops with increasing gas pressure and therefore density [13] due to the decrease of the electron impact ionisation yield in the reduced electric field (E/P) within the avalanche region, and the limit due to breakdown imposed on the potential difference between the electrodes.

Although the breakdown limit of a TGEM operating in charge gain mode in pure noble gas is ultimately determined by the effect of UV photon emission, in a heavily optimised system gain is limited by charge build up within the amplification region. The high electric fields required for charge amplification can also lead to charging of the dielectric and ultimately to dielectric breakdown. Additionally, high levels of UV photon emission (secondary scintillation) always associated with charge gain, reduce the device stability, and bubble formation on the sharp edges of the TGEM electrodes can lead to the creation of conducting paths culminating in gaseous discharge.

Secondary scintillation was only observed in gas at TGEM fields for which charge amplification was already established, with the effect that ionisation electrons created in the charge multiplication process then interacting with the medium, yield additional secondary scintillation and therefore an exponential scaling with the applied field. No evidence of charge multiplication was detected during this work for any electric field within liquid argon and it is assumed that the field required to initiate secondary scintillation via excitation is below the ionisation threshold of the medium, the energy dissipated via the emission of UV photons, which do not participate further in the process. Secondary scintillation therefore scales linearly with electric field in the liquid, and may therefore eventually be considered more predictable and less prone to breakdown. Results contained within this report have demonstrated that using secondary scintillation as the readout signal, it is possible to achieve high gains and potentially satisfactory energy resolution from a single TGEM in liquid argon. By operating the TGEM at a lower field in the secondary scintillation proportional mode, detector instability is reduced, and dielectric and UV photon mediated breakdown are avoided. Readout of secondary scintillation also increases signal to noise ratio and reduces the requirements of the front end electronics on the TGEM.

5. Conclusions

For the first time secondary scintillation, generated within the holes of a thick gas electron multiplier (TGEM) immersed in liquid argon, has been observed and measured using a silicon photomultiplier device (SiPM). Secondary scintillation has also been observed in 1 bar cold saturated vapour in a double phase chamber. In both cases the origin of the photon production was ascertained by reduction of the drift field. In order to explain the exponential nature of secondary scintillation generation, measurements of charge gain were taken in gaseous argon, comparison of the voltage range across the TGEM for each process revealing the resulting scintillation gain to be a combination of both.

The combination of the waveshifter tetraphenyl butadiene (TPB) within a gel applied to the surface of the SiPM device was found to efficiently absorb 128nm VUV light produced within the TGEM holes, and to then emit 460nm in the high quantum efficiency region of the device. Overall, for a SiPM over-voltage of 1V, a TGEM voltage of 9.91KV, and a drift field of 2.5KV/cm, a total of 62 ± 20 photoelectrons were produced at the SiPM device per Fe-55 event, corresponding to an estimated gain of 150 ± 66 photoelectrons per drifted electron.

This new liquid argon detection technology is considered to hold great potential on the route to cost-effective, large volume, simultaneous tracking and calorimetry targets with excellent performance relevant to important applications in fundamental particle physics.

References

[1] F. Sauli, *GEM: A new concept for electron amplification in gas detectors*, Nucl. Instr. and Meth. **A386**, (1997), 531.

[2] A. Bondar, A. Buzulutskov, A. Grebenuk, D. Pavlyuchenko, R. Snopkov, Y. Tikhonov, V.A. Kudryavtsev, P.K. Lightfoot, N.J.C. Spooner, *A two-phase argon avalanche detector operated in a single electron counting mode*, Nucl. Instr. and Meth. **A574**, (2007), 493.

[3] P.K.Lightfoot, et al., *Development of a double-phase xenon cell using micromegas charge readout for applications in dark matter physics*, Nucl. Instr. and Meth. **A554**, (2005), 266.

[4] P.K. Lightfoot, et al., *First operation of bulk micromegas in low pressure negative ion drift gas mixtures for dark matter searches*, Astropart. Phys. **27**, (2007), 490.

[5] J.G.Kim, et al., *Electron avalanches in liquid argon mixtures*, Nucl. Instr. and Meth. **A534**, (2004), 376.

[6] P.K. Lightfoot, et al., *Characterisation of a silicon photomultiplier device for applications in liquid argon based neutrino physics and dark matter searches*, **JINST 3**, (2008), P10001.

[7] K. Mavrokoridis, et al., *Development of wavelength shifter coated reflectors for the ArDM argon dark matter detector*, awaiting submission to JINST.

[8] M. Miyajima, et al., *Average energy expended per ion pair in liquid argon*, Phys. Rev. **A9**, (1974), 1438.

[9] A. Hitachi, et al., *Effect of ionization density on the time dependence of luminescence from liquid argon and xenon*, Phys. Rev **B 27**, (1983), 5279.

[10] J.S. Kapustinsky, R.M. DeVries, N.J. DiGiacomo, W.E. Sondheim, J.W. Sunier, H. Coombes, *A fast timing light pulser for scintillation detectors*, Nucl. Instr. and Meth. **A241**, (1985), 612-613.

[11] A. Bondar, et al., *Two phase argon and xenon avalanche detectors based on gas electron multipliers*, Nucl. Instr. and Meth. **A556,** (2006), 273.

[12] S.E. Derenzo, T.S. Mast, R.A. Muller, and H. Zaklad, *Electron avalanche in liquid xenon*, Phys. Rev. A, vol. **9**, (1974), 2582.

[13] F.D. Amaro, et al., *Operation of a single-GEM in noble gases at high pressure*, Nucl. Instr. and Meth. **A579**, (2007), 62.

An ultra-low-background detector for axion searches

S Aune[1], T Dafni[2], G Fanourakis[3], E Ferrer Ribas[1], J Galán Lacarra[2], T Geralis[3], I Giomataris[1], F J Iguaz[2], I G Irastorza[2], K Kousouris[3], J Morales[2], J P Mols[1], T Papaevangelou[1,5], J Ruz[2], A Tomás[2] and T Vafeiadis[4]

[1]IRFU, Centre d'Etudes de Saclay, Gif sur Yvette CEDEX, France

[2]Instituto de Física Nuclear y Altas Energías, Zaragoza, Spain

[3]Institute of Nuclear Physics, NCSR Demokritos, Athens, Greece

[4]Physics Department, Aristotle University, Thessaloniki, Greece

[5]E-mail: Thomas.Papaevangelou@cern.ch

Abstract. A low background Micromegas detector has been operating in the CAST experiment at CERN for the search of solar axions since the start of data taking in 2002. The detector, made out of low radioactivity materials, operated efficiently and achieved a very low level of background (5×10^{-5} keV^{-1}·cm^{-2}·s^{-1}) without any shielding. New manufacturing techniques (Bulk/Microbulk) have led to further improvement of the characteristics of the detector such as uniformity, stability and energy resolution. These characteristics, the implementation of passive shielding and the improvement of the analysis algorithms have dramatically reduced the background level (2×10^{-7} keV^{-1}·cm^{-2}·s^{-1}), improving thus the overall sensitivity of the experiment and opening new possibilities for future searches.

1. Introduction

The CAST (CERN Axion Solar Telescope) experiment is using a decommissioned LHC dipole magnet to convert solar axions into detectable X-ray photons. Axions are light pseudoscalar particles that arise in the context of the Peccei-Quinn [1] solution to the strong CP problem and can be Dark Matter candidates [2]. Stars could produce axions via the Primakoff conversion of the plasma photons. The CAST experiment is pointing at our closest star, the Sun, aiming to detect solar axions. The detection principle is based on the coupling of an incoming axion to a virtual photon provided by the transverse field of an intense dipole magnet, being transformed into a real, detectable photon that carries the energy and the momentum of the original axion. The axion to photon conversion probability is proportional to the square of the transverse field of the magnet and to the active length of the magnet. Using an LHC magnet (9 T and 9.26 m long) improves the sensitivity by a factor 100 compared to previous experiments. The CAST experiment has been taking data since 2003 providing the most restrictive limits on the axion-photon coupling [3]-[6] for masses $m_a < 0.02$ eV. At this mass the sensitivity is degraded due to coherence loss. In order to restore coherence, the magnet can be filled with a buffer gas providing an effective mass to the photon [4]. By changing the density of the buffer gas in steps, one can scan a wide range of axion mass values. At the end of 2005 CAST started such a program, entering its phase II by filling the magnet bore with He gas. From 2005 to 2007, the magnet bore was filled with ^{4}He gas extending the sensitivity to masses up to 0.4 eV [7]. From March

2008 onwards the magnet bore has been filled with ^{3}He and the sensitivity should be increased up to $m_a < 1.2$ eV by the end of the ^{3}He run in 2010.

The CAST experiment has used three different types of detectors to detect the X-rays originated from the conversion of the axions inside the magnet: a Time Projection Chamber [7], an X-ray telescope [9], and a Micromegas detector [10]. The Micromegas detector of CAST is a gaseous detector optimized for the detection of low energy (1–10 keV) X-ray photons. It is based on the micropattern detector technology of Micromegas (MICROMEsh GAseous Structure) developed in the mid 90's [11]-[13]. The advantages of using Micromegas for such low-threshold, low-background measurements as required by the CAST experiment, include sensitivity at the keV and sub keV energy region where very good energy resolution can be achieved, excellent spatial resolution, one dimensional or X-Y readout capability, stability, construction simplicity and low cost. In addition it allows flexibility in the construction materials, the proper choice of which would lead to a detector appropriate for low background measurements.

2. The Micromegas detector in CAST

During CAST phase I and ^{4}He data taking periods, a conventional Micromegas was used [10], covering one of the 4 magnet bore exits. Thought it was the only detector without any shielding it showed the lowest background level of $\sim 5\times10^{-5}$ keV$^{-1}\cdot$cm$^{-2}\cdot$s^{-1} of all 3 detectors used. Due to this fact and, in combination with the development of new manufacturing techniques (Bulk & Microbulk, [14]), the existing Micromegas and TPC were replaced with three new Micromegas detectors for the ^{3}He phase. The compact design of the new detectors allowed the implementation of passive shielding, which further reduced the background by a factor ~3, reaching the level of $\sim 1\times10^{-5}$ keV$^{-1}\cdot$cm$^{-2}\cdot$s^{-1} [15] during April – September 2008 data taking period.

In the following cryogenic maintenance intervention two of these detectors were replaced by newer Microbulk type models. These detectors are expected to have a very low intrinsic background since they are made of low radioactivity materials: kapton and copper for the detector and Plexiglas and aluminum for the chamber. Radiopurity measurements of several detectors parts are being done in the Canfranc Underground Laboratory, and are used to continuously improve the intrinsic radioactivity of new series of detectors and to build a full simulation model for the detector background that will certainly be of great use to study the experimental results presented in this paper.

2.1. Background reduction

The data acquisition system used in CAST Micromegas is registering the analogue signal from the mesh with 1 GHz FADC and the integrated charge on each strip using Gasiplex cards [10]. From this information several parameters can be extracted, that characterize the interaction of particles with the detector gas, like risetime and pulse duration from the analogue signal and hit multiplicities from the strip data. An X-ray of energy less than 10 keV would cause primary ionization localized in a range less than 1 mm. This would give a narrow pulse with a fixed risetime determined by the detector characteristics, and a mean strip multiplicity which corresponds to ~5 mm. A minimum ionizing particle like cosmic muons would cause spatially extended ionization resulting to broader pulses and higher multiplicities (Figure 1).

The raw trigger rate of the Micromegas detectors in CAST is of the order of 1 Hz. Most of these events concern muons and can be rejected in the offline analysis with the proper particle recognition algorithm. The aim of such algorithms is to reject all signals that are not compatible with X-rays, but maintaining most of the X-ray compatible ones. The data reduction factor has reached the level of 10^{-3}, while maintaining 80% of the 6 keV X-rays from ^{55}Fe calibration runs. The low level of the non shielded Micromegas detector used in phase I has been achieved by only this data reduction.

2.2. Detector shielding

Environmental gamma rays and neutrons could cause X-ray compatible signals in the detector via Compton and elastic scattering respectively. If the deposited energy in the gas is of the order of few

Journal of Physics: Conference Series **179** (2009) 012015 doi:10.1088/1742-6596/179/1/012015

keV the signal will have characteristics similar to the X-rays and can not be rejected offline. These signals can be only reduced by the implementation of proper shielding.

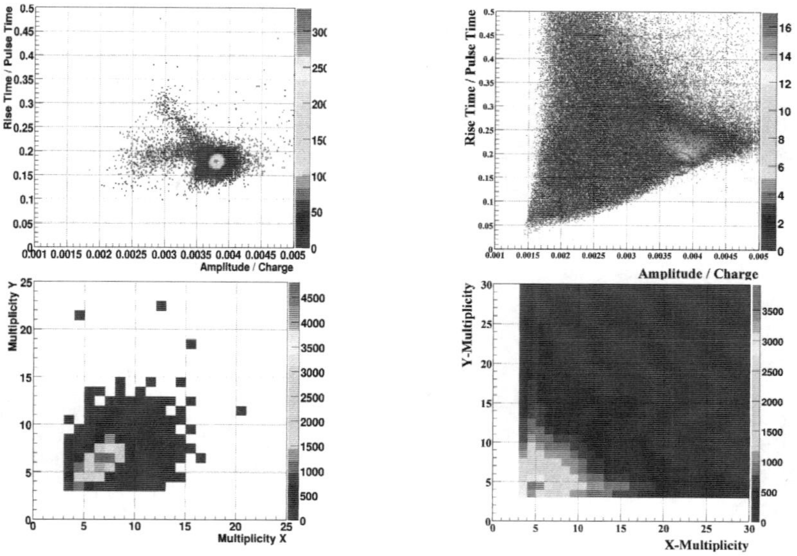

Figure 1. Up-left: correlation of mesh pulse characteristics for a calibration run with 6 keV X-rays from ^{55}Fe. Up-right: the same parameters for a data (cosmics) run. Down left: strip multiplicities for a calibration run. Down right: strip multiplicities for a data run. By applying appropriate cuts the cosmic data can be reduced by a factor of ~10^{-3}, while the X-rays are reduced only by ~20%.

The shielding that was adopted for the Micromegas detectors consists of an inner archeological Pb layer 3-5 cm thick, and an external polyethylene layer 15-20 cm thick. In-between a Cd foil is placed in order to absorb the neutrons which have been thermalized in the polyethylene. A 5 mm Cu layer is placed inside the lead serving also as faraday cage. The whole shielding must be made air-tight and is flushed with nitrogen in order to remove Radon. Studies done in the CAST experimental area have shown that such shielding results to a background reduction of a factor more than 3 (Figure 2).

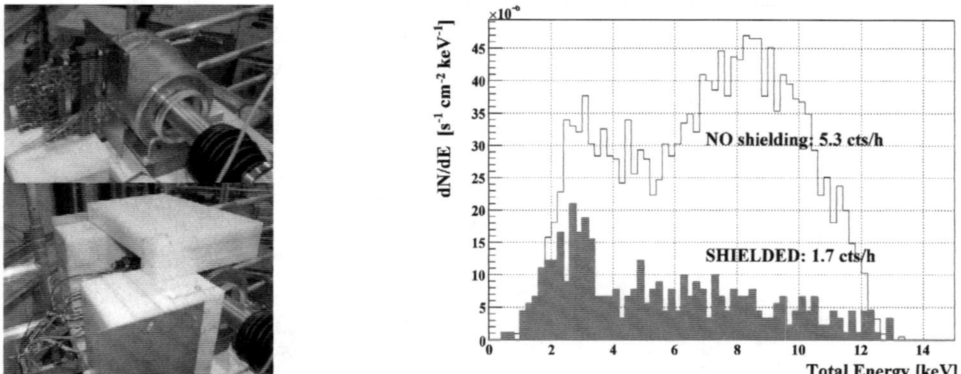

Figure 2. Left: pictures of the CAST sunrise Micromegas shielding. In the ideal case the detector should be covered by all shielding elements in a 4π geometry. However this was not possible due to space limitations and detector electronics. Right: results from the background reduction tests. A reduction more than 3 is observed in the 1-7 keV energy range which was used for the limit calculation in phase I. This reduction is bigger in higher energies.

3. The "Ultra-low-background" case

3.1. The measurements

Special care was taken for the shield air-tightness during the installation of the new detectors in October 2008. Since then, the detector was taking data in stable conditions for a period over than one month. ^{55}Fe calibration runs were taken regularly (~daily) at random times during the day, in order to check the gain and software cuts stability. The source was placed in a pneumatic mechanism that could move it inside the CAST vacuum tube in front of the Micromegas window at a distance of 1m, thus simulating X-rays that would emanate from the axion to photon conversions in the 9T magnetic field. These runs have shown a stable detector performance (Figure 3).

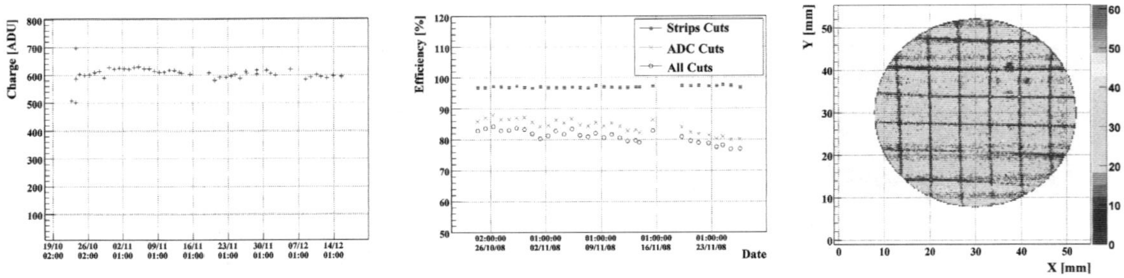

Figure 3. ^{55}Fe calibration runs during Oct-Nov 2009. The gain stability is shown on the left plot, and the software efficiency on the middle one. The right plot is showing the x-y image of the calibration runs in the area of expected signals. The image of the drift window strongback is clearly visible, while no serious defect is seen in the whole detector area.

The background level of this detector was initially similar to the one of the previous detectors used at the same place with similar shielding, but it appeared to decrease with time (Figure 4). The decrease rate was compatible with Radon's decay time, which could be an indication that the observed background was dominated by Radon radioactivity. This reduction stopped on the 4th of November and the background level returned to the initial values. It was found later that the moment of the increase coincided (within an hour) with an intervention on the nitrogen central supply line which caused interruption of the shielding nitrogen flushing.

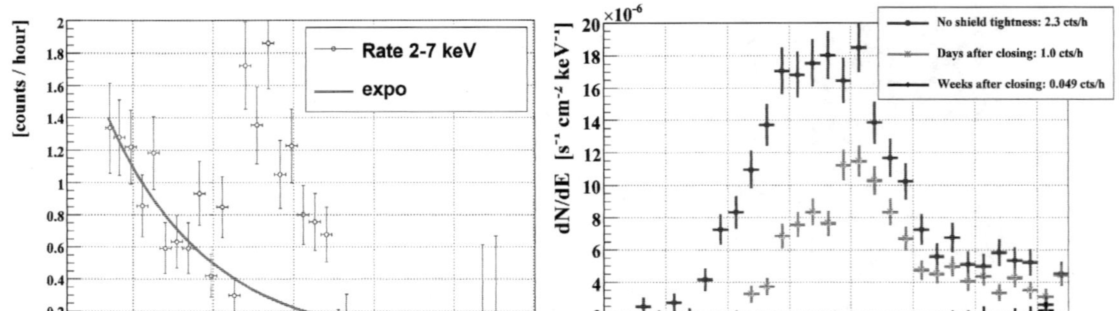

Figure 4. Left: background level evolution with time; the sudden increase on the 4/11 coincided with an interruption of the nitrogen flushing. Right: evolution of the analogue spectrum; the three spectra correspond to the data of first few days (blue), the second week (red) and the last two weeks of (black)

The background rate started decreasing again after the interruption and a week later reached a surprisingly low level of the order of 2×10^{-7} keV$^{-1}\cdot$cm$^{-2}\cdot$s^{-1}. It remained at this level until the 27th of

November when a series of power cuts lead to the end of the run. During this period of two weeks the background level of data runs was stable at this very low level, while no defect has been observed in the calibration runs.

3.2. Discussion

Several considerations about this data taking period allow us to exclude that the observed ultra-low-background is due to a detector malfunction or an artefact.

The ^{55}Fe calibration runs taken during this period show that the detector response to 6 keV X-rays remained practically unchanged, as shown in Figure 3, while the reduction is observed for that energy as well (Figure 4). Furthermore, the fact that the calibrations were taken at random moments during the day exclude possible diurnal effects.

No hardware problems that would imply detection defects like increased deadtime have been observed. A characteristic example of a long run is presented in Figure 5, where the detector trigger rate appears to be stable in the nominal value (0.7 Hz) during 3 days, while only 5 events remain after the software cuts. Similar behavior was observed in the (shortest) runs of this period as well.

The analogue spectrum after cuts seems to remain unchanged in shape and it only scales down with time (Figure 4). This fact is compatible with the assumption that most of the observed counts are fluorescence X-rays caused by external radiation. Since the shielding is stopping most of the environmental gammas and the detector consists of low radioactivity materials the only remaining source of external radiation is radon trapped inside the shielding. In this case the lines of detector's and shielding copper (8 keV), tube's stainless steel iron (6 keV) and window's aluminum (1.5 keV) would be excited and the accompanying argon escape lines (E_x-3keV) would also appear. The intensity of these lines should decrease with time if the shielding is air-tight. This kind of behavior is observed in Figure 4.

The radon scenario is also supported by the interruption of the nitrogen flushing and the simultaneous increase of the background rate.

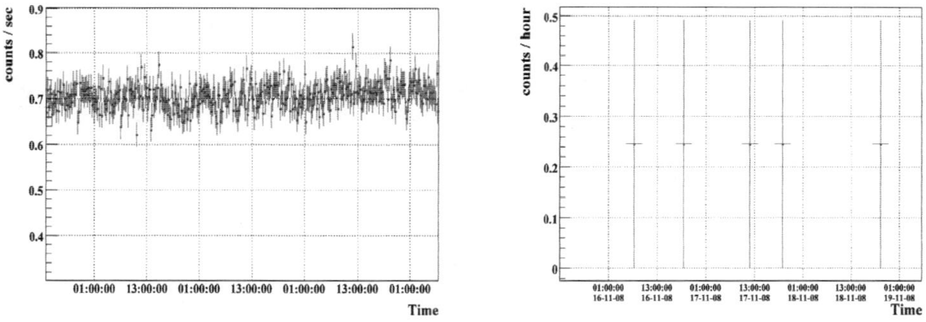

Figure 5. Raw trigger rate during a three day long run (left) and X-ray compatible events after cuts (right). The trigger rate was very stable during the run, while only 5 X-ray compatible events remained after applying the analysis cuts.

Further analysis of data taken during this period has revealed another characteristic of the detector's behavior. Though the X-ray pulse parameters remained unchanged during the whole period, some of the parameters of the cosmic runs are changing, especially during the ultra low background period. This fact leads to the conclusion that the assumed "radon effect" could be observed in combination with an enhanced particle recognition efficiency. The reason of such behavior is currently under investigation with several detectors and configurations.

Similar behavior has been observed with the second detector as well, which was installed for testing at the same shielding for approximately 10 days during the shutdown. In this period the count rate dropped exponentially from the initial 2 to 0.3 counts per hour.

4. Conclusions

Rare event detection experiments, such as search for solar axions with CAST, demand stable and low background detectors. These demands are met by the Micromegas detector which can be constructed by low radioactivity materials and has properties that allow high background rejection efficiency. Such detectors have been used in CAST through all data taking campaigns. The implementation of passive shielding has enhanced the performance of these detectors, meeting the needs of CAST phase II, which demands a background level of the order of zero counts per hour. The recently developed Microbulk Micromegas lead to further reduced background levels. Its enhanced performance, that would imply 0.05 expected background counts per hour in the axion expected signal energy region, has been verified during on situ test measurements.

The observed ultra-low-background level not only simplifies the CAST ^{3}He data taking phase, but also opens new opportunities for future improvements. As an example can be noted the phase I result, which represents the strongest experimental bound in axion coupling constant [3]. This result has been dominated by the X-ray telescope [9], which showed a background rate of 0.15 counts per hour. Repeating these measurements using three new Micromegas detectors in parallel with the telescope, would lead to an improvement of the coupling constant limit by almost a factor of two, probing for the first time the axion parameter space well beyond astrophysical bounds [18].

References

[1] R D and Quinn H R, Phys. Rev. Lett. 38, 1440 (1977).

[2] Sikivie P, Int. J. Mod. Phys. D3 S, 1 (1994) and Phys. Rev. Lett. 51, 1415 (1983).

[3] Andriamonje Set al. (CAST collaboration), "An improved limit on the axion-photon coupling from the CAST experiment", 2007 JCAP 04, 10 [SPIRES] [hep-ex/0702006].

[4] Van Bibber K, McIntyre P M, Morris D E and Raffelt G G, "A practical laboratory detector for solar axions", 1989 Phys. Rev. D 39, 2089.

[5] Zioutas K et al, "A decommissioned LHC model magnet as an axion telescope", 1999 Nucl. Instrum. Methods Phys. Res., A 425 480 [astro/ph-9801176].

[6] Zioutas K et al. (CAST collaboration), "First results from the CERN Axion Solar Telescope", 2005 Phys. Rev. Lett. 94, 121301 [SPIRES] [hep-ex/0411033].

[7] Arik E et al. (CAST collaboration), "Probing the Axion models with CAST", JCAP 0902.008 (2009), [arXiv:0810.4482]

[8] Autiero Det al., "The CAST Time Projection Chamber", 2007 New J. Phys 9, 171 [SPIRES] [hep-ex/0702189].

[9] Kuster M et al.,"The X-ray telescope of CAST", 2007 New J. Phys 9, 169 [SPIRES] [hep-ex/0702188].

[10] Abbon P et al., "The Micromegas detector of the CAST experiment", 2007 New J. Phys 9, 170 [SPIRES] [hep-ex/0702190].

[11] Giomataris Y, Rebourgeard P, Robert J P & Charpak G 1996 Nucl. Instrum. Methods Phys. Res., Sect. A 376, 29–35.

[12] Giomataris Y 1998 Nucl. Instrum. Methods Phys. Res., Sect. A 419, 239–250.

[13] Charpak G, Derr´e J, Giomataris Y & Rebourgeard P 2002 Nucl. Instrum. Methods Phys. Res., Sect.A 478, 26–36.

[14] Giomataris Y et al., "Micromegas in a bulk", Nucl. Instrum. Meth. A 560 2, 2006

[15] S. Aune et. al. "New Micromegas detectors in the CAST experiment", Nucl. Instrum. Meth. A 604 Issues 1-2, 2009

[16] Niinikoski T O et al., "Thin cryogenic X-ray windows", 2008 Submitted to Proc. ICEC 22, Seoul.

[17] CAST Collaboration, "Technical Design Report of the CAST 3He Gas System", SPSC-TDR-001, CERN-SPSC-2006-029.

[18] T. Papaevangelou, presentation at "New Opportunities in the Physics Landscape at CERN" workshop, CERN, 10-13 May 2009.

Relic Neutrino Detection Using Neutrino Capture on Beta Decaying Nuclei

Alfredo G. Cocco

Istituto Nazionale di Fisica Nucleare - Sezione di Napoli
Complesso Universitario di Monte S.Angelo - I80126 Napoli (Italy)

E-mail: alfredo.cocco@na.infn.it

Abstract. In this paper we present a study of the interaction of low energy electron antineutrino on nuclei that undergo electronic capture. This completes the investigation on the possibility to detect very low energy neutrino and antineutrino using unstable nuclei that has been recently addressed by the study of neutrino capture on $\beta^{\pm}$ decaying nuclei. Experimental challenges toward the detection of Cosmological Relic Neutrino Background by the use of these reactions are discussed.

1. Introduction

In a previous paper [1] we have shown that nuclei that decay via electron or positron emission are suitable to detect very low energy neutrino. This indeed appears to be a general property of all nuclear decays in which a neutrino (or antineutrino) is present in the final state: due to particle crossing from the final to the initial state the resulting reaction has no threshold on the energy of the incoming (anti)neutrino. The fact that neutrino has a non zero mass can have important consequences on the kinematics of the capture process: in the case of beta decay treated in [1] this led to the possibility to unambiguously detect neutrino of arbitrarily low energy.

The case of nuclei that decay through the electron capture (EC) process appear to be worth to explore as well; in the EC process the nucleus of a neutral atom A captures one bound electron to produce a daughter atom B and an electron neutrino

$$e^- + A^+ \rightarrow B^* + \nu_e \rightarrow B + \nu_e + n\gamma \ . \qquad (1)$$

The atom B, initially in an excited state having an electron vacancy in an inner atomic shell, decays electromagnetically releasing a total energy E_l; by simple considerations it turns out that E_l is the captured electron binding energy in the field of the daughter nucleus. The nucleus of atom B can be produced in an excited nuclear state as well.

Electron capture transitions are suitable to detect electron antineutrino via two crossed reactions: the first is given by

$$\bar{\nu}_e + A \rightarrow B^- + e^+ \qquad (2)$$

and involves a simultaneous exchange of the electron and the neutrino in the initial and final states. The second reaction that can be obtained by crossing from the EC decay is given by

$$\bar{\nu}_e + e^- + A^+ \rightarrow B \qquad (3)$$

which involves the exchange of the electron neutrino in the final state to an electron antineutrino in the initial state. In the following we will show under which circumstances these two process could be used for the detection of very low energy antineutrino.

2. Kinematics of antineutrino capture on EC decaying nuclei

The behavior of reaction 2 as a function of the Q-value can be divided in three categories: in case $Q_{EC} > 2m_e - m_\nu$ the neutrino capture process of reaction 2 has indeed no energy threshold since the Q-value is large enough to allow the creation of a positron in the final state even without the contribution of the electron mass in the initial state of the EC decay. On the other hand, if the Q-value satisfy the relation $Q_{EC} > 2m_e + m_\nu$ positron decay become energetically allowed; this case is described in details in [1] and will not be treated here. There exist thus a range of values of Q that is $2m_\nu$ wide and that would allow the detection of antineutrino with an arbitrarily small energy. Transitions falling in this category would have the remarkable property of a unique signature: the positron in the final state of reaction 2 can be definitely used to disentangle the antineutrino capture interaction with respect to the spontaneously occurring EC decay. Finally, in the case of antineutrino captured on nuclei having $Q_{EC} < 2m_e - m_\nu$ reaction 2 presents a threshold on the energy of the incoming antineutrino given by $E_\nu^{\mathrm{thr}} = 2m_e - Q_{EC}$ due to the fact that the Q-value cannot compensate anymore the lack of the captured electron mass in the initial state and the creation of a positron in the final state. The energy of the outgoing positron is given in fact by $E_e = E_\nu + Q_{EC} - m_e$.

For what concerns reaction 3 the Q-value plays a crucial role as well. In case $Q_{EC} - E_l \geq -m_\nu$ reaction 3 has no energy threshold; moreover, the nucleus A is stable since the corresponding EC decay become energetically allowed only if $Q_{EC} - E_l \geq m_\nu$. Thus, once again, it exists a region of Q-values that is $2m_\nu$ wide and in which reaction 3 has no threshold on the energy of the incoming antineutrino. The process is also background free since the EC decay is energetically forbidden. Nevertheless, reaction 3, as it is, is forbidden by the lack of a suitable final state. Using the second Fermi golden rule ($w = 2\pi/h\,|\mathcal{M}|^2\,\rho_f$) it is easy to prove in fact that the cross section for this process depends on the number of available final states per unit energy $\rho(E_\nu) = \delta(E_\nu - (Q_{EC} - E_l))$; this has only one possible solution in case of antineutrino at rest and $Q_{EC} - E_l = -m_\nu$. Despite that, we can still envisage at least two other cases in which the process might happen: 1) there exists an excited state B' having energy $M(B) + E_\nu + Q_{EC} - E_l$; in this case EC decay through the same channel would be forbidden due to energy conservation even in the limit of $E_\nu \to m_\nu$ for non vanishing values of the neutrino mass; 2) the captured electron is "off-mass shell" with an effective mass given by $m_{\mathrm{eff}} = m_e - Q_{EC} + E_l - E_\nu$ and this could happen for example in a metal when the nucleus captures an electron in the valence band, being in this case E_l the mean binding energy of valence electrons. In the case of $Q_{EC} - E_l < -m_\nu$ reaction 3 could still be triggered by antineutrino having an energy greater than $E_\nu^{\mathrm{thr}} = -Q_{EC} + E_l$.

3. Antineutrino capture cross section

The electron capture process has been studied in detail in [2]. The rate of electron capture is given by

$$\lambda_{EC} = \frac{G_\beta^2}{2\pi^3} \sum_x n_x C_x(p_e, p_\nu) f_x \tag{4}$$

where the sum extends over all the atomic shells from which an electron can be captured, n_x is the relative occupation number of that shell and $C_x(p_e, p_\nu)$ is the nuclear shape factor relative to the given transition. The index x labels the orbital electron wave-function via the variable κ_x given by the spherical waves decomposition, as described in [2]. The function f_x is the analogous of the integrated Fermi function of the β decay and is given by the following expression

$$f_x = \frac{\pi}{2} q_x^2 \beta_x^2 B_x \tag{5}$$

where $q_x = (Q_{EC} - E_l)/m_e$ is the neutrino energy, β_x is the Coulomb amplitude of the bound-state electron radial wave-function and B_x is the associated electron exchange and overlap correction.

In complete analogy with the procedure used in [1] the antineutrino capture cross section for reaction 2 can be written as

$$\sigma_{\rm NCB}(E_\nu)v = \frac{G_\beta^2}{\pi} p_e E_e F(Z, E_e) C(p_e, p_\nu)_\nu \tag{6}$$

where $C(p_e, p_\nu)_\nu$ is the shape factor of the antineutrino capture interaction and should be evaluated using prescription given in [3] and $F(Z, E_e)$ is the Fermi function to be evaluated for the outgoing positron. It is worth notice here that $C(p_e, p_\nu)_\nu$ and $C_x(p_e, p_\nu)$ contain the same nuclear form factors; moreover, $C(p_e, p_\nu)_\nu$ can be expressed as a weighted sum of the $C_x(p_e, p_\nu)$ provided the correct superposition of free-particle and bound state electron wave-functions are evaluated for each specific case. We will show that in most of the cases $C(p_e, p_\nu)_\nu$ can be obtained to a very good approximation using leading order $C_x(p_e, p_\nu)$ terms.

In the case of reaction 2 it is possible to use expression 4 to rewrite the cross section 6 in the following way

$$\sigma_{\rm NCB}\, v = 2\pi^2 \ln 2\; p_e E_e \frac{F(Z, E_e) C(p_e, p_\nu)_\nu}{t_{1/2}^{\rm EC}\, \sum_x n_x C_x f_x}. \tag{7}$$

According to the procedure used in [1] we define a shape factor ratio $\mathcal{A}$ as

$$\mathcal{A} = \frac{\sum_x n_x C_x(q_\nu) f_x(q_\nu)}{p_e E_e F(Z, E_e) C(p_e, p_\nu)_\nu} \tag{8}$$

where q_ν is the energy of the outgoing neutrino in the EC decay and the subscript (e) refers to the positron in the final state of reaction 2. Using this position the antineutrino cross section can be written as

$$\sigma_{\rm NCB}\, v = \frac{2\pi^2 \ln 2}{\mathcal{A} \cdot t_{1/2}^{\rm EC}}. \tag{9}$$

In case of reaction 3 it is easy to prove that the antineutrino cross section is given by

$$\sigma_{\rm NCB}\, v = \frac{G_\beta^2}{\pi} \sum_x n_x C_x(p_e, p_\nu)_\nu \rho_x(E) g_x \tag{10}$$

where $\rho(E)$ is the number of available final states per unit energy and

$$g_x = \frac{\pi}{2} \beta_x^2 B_x \tag{11}$$

is written in analogy to expression 5. In the case in which all energies in the initial and final states are given, $\rho(E)$ is a Dirac delta function $\delta(E_\nu + Q_{EC} - E_l)$.

In the case of reaction 3 the shape factor ratio is given by

$$\mathcal{A}' = \frac{\sum_x n_x C_x(q_\nu) f_x(q_\nu)}{\sum_x n_x C_x(E_\nu) g_x \rho_x(E_\nu)} \tag{12}$$

where the variable q_ν refers to the neutrino energy in the final state of the EC process while E_ν is the energy of the incoming neutrino in reaction 3; also in this case the cross section can be written according to expression 9.

It is easy to prove that the shape factor ratios $\mathcal{A}$ and $\mathcal{A}'$ can be evaluated to a very good approximation for allowed decays and in an exact way for superallowed and forbidden unique decays; a detailed discussion is reported in [4].

Table 1. Pure EC decaying nuclei that present the largest product of $\sigma_{\rm NCB} \cdot t^{\rm EC}_{1/2}$ for the process in expression 2. Cross section is evaluated for incoming antineutrino energy of 1 MeV above reaction threshold and in case of K shell capture. Allowed transitions (top) and forbidden unique (bottom) are shown.

Isotope	Decay $(J_i \rightarrow J_f)$	$E_\nu^{\rm thr}$ (keV)	Half-life (sec)	$\sigma_{\rm NCB}$ $(10^{-41}$ cm$^2)$
^{7}Be	$\frac{3}{2}^- \rightarrow \frac{1}{2}^-$	637.80	4.40×10^7	6.80×10^{-3}
^{7}Be	$\frac{3}{2}^- \rightarrow \frac{3}{2}^-$	160.18	5.13×10^6	1.16×10^{-2}
^{55}Fe	$\frac{3}{2}^- \rightarrow \frac{5}{2}^-$	790.62	8.64×10^7	1.55×10^{-5}
^{68}Ge	$0^+ \rightarrow 1^+$	916.00	2.34×10^7	1.39×10^{-4}
^{178}W	$0^+ \rightarrow 1^+$	930.70	1.87×10^6	5.14×10^{-4}
^{41}Ca	$\frac{7}{2}^- \rightarrow \frac{3}{2}^+$	600.61	3.22×10^{12}	8.35×10^{-9}
^{81}Kr	$\frac{7}{2}^+ \rightarrow \frac{3}{2}^-$	741.30	7.23×10^{12}	2.40×10^{-9}
^{100}Pd	$0^+ \rightarrow 2^-$	693.68	3.14×10^5	4.17×10^{-4}
^{123}Te	$\frac{1}{2}^+ \rightarrow \frac{7}{2}^+$	970.70	1.89×10^{22}	5.40×10^{-15}

4. Estimating antineutrino capture cross section

The cross section for reaction 2 can be evaluated using expression 9 and following the procedure illustrated in the section II. Numerical values for the constants appearing in expression 5 can be found in [2]. A description of the algorithm used to compute the Fermi function is given in [1]. The values of the cross section for nuclei having the largest product of cross section times lifetime for a specific value of the incoming neutrino energy are reported in table 1.

The antineutrino cross section for reaction 3 can be easily written in the case of electron captured from the K shell

$$\sigma_{\rm NCB}\, v = \frac{2\pi^2 \ln 2}{q_\nu^2 \; t^{\rm EC}_{1/2}} \rho(E_\nu) \tag{13}$$

where q_ν and $t^{\rm EC}_{1/2}$ are, respectively, the outgoing neutrino momentum and the half–life of the nucleus EC decay.

5. Antineutrino capture versus EC decay rate

In the case of reactions 2 and 3 and using, respectively, equations 8 and 12 we can write the ratio between antineutrino capture and EC decay rate as

$$\frac{\lambda_\nu}{\lambda_{EC}} = \frac{2\pi^2}{\mathcal{A}^{(\prime)}} n_{\bar\nu} \tag{14}$$

where $n_{\bar\nu}$ is the antinuetrino density at the nucleus. As an example, we have that the ratios

$$\frac{\lambda_\nu}{\lambda_{EC}} = 4\pi \frac{n_{\bar\nu}}{\sum n_x \beta_x^2 B_x} \frac{p_e E_e F(Z, E_e)}{q_{\bar\nu}^2} \quad \text{and} \quad \frac{\lambda_\nu}{\lambda_{EC}} = 2\pi^2 \frac{n_{\bar\nu} \rho(E_\nu)}{q_{\bar\nu}^2} \tag{15}$$

represent, respectively, the case of reaction 2 for superallowed transitions and case of reaction 3 when considering only K-shell capture in superallowed and K-th unique transitions.

6. Cosmological relic antineutrino detection

The detection of the cosmological relic neutrino background (CνB) using reaction 2 is of course problematic due to the presence of the energy threshold; neutrino of the CνB have an energy of the order of the neutrino mass and thus in the electron-volt scale. A possible solution could be to use the CνB as a target for accelerated nuclei; in this case the threshold energy could be accomplished by the center of mass energy of the accelerated nucleus and the steady neutrino. From simple kinematical considerations we have that the nucleus of mass M should have a γ greater than $\gamma_{\rm min} = E_{\rm thr}/m_\nu$. The relic antineutrino capture rate can thus be expressed as

$$\lambda_{\rm NCB} = \frac{n_{\bar\nu}\, 2\pi^2 \ln 2}{\mathcal{A} \cdot t^{\rm EC}_{1/2}} \tag{16}$$

where $t^{\rm EC}_{1/2}$ is the decay's proper time. The total rate is obtained by multiplying the last expression by the total number of accelerated nuclei $\mathcal{N}$ where realistic values for present storage rings are $\mathcal{N} = 10^{13}$. Assuming a transition having a value of $Q_{\rm eff}$ of the order of the elctronvolt this would lead to an interaction rate in case of allowed transitions of the order of $\lambda_{\rm NCB} \simeq 10^{-18}$ s^{-1}, too slow to be effectively detected even in case of absence of background due to EC decay of the nucleus (in case of a fully ionized beam).

The CνB detection using reaction 3 appears even more difficult due to the fact that in the case of neutrino having very small energy the number of final states per unit energy $\rho(E)$ is basically unknown. The atom in the final state has to accomplish an excess energy $Q_{EC} - E_l + m_\nu$ and this can happen only if this energy can be radiated out via electromagnetic or phonon emission (if the decaying atom is bounded in a solid). Photons emission can be due either to atomic electrons or to nuclear level transition; in the first case the typical energy lies in the eV-keV region and, being E_l in the same energy range, this impose that only nuclei with a very small Q-value could be suitable for this detection. In the second case there should exist a nuclear level that matches the energy difference. In both cases the energy of these levels should match exactly the mass difference; moreover, to avoid the possibility that the channel opens up also for the spontaneous EC decay this level must be m_ν above the transition Q-value. On the other hand, being in this case the EC decay forbidden, there is a priori no easy way to evaluate the cross section for reaction 3.

7. Conclusions

The analysis of antineutrino interactions on nuclei that decay via Electronic Capture shows that the two crossed reactions that could be obtained have a sizable cross section and are both suitable to detect low energy antineutrino. In spite of that, the presence of an energy threshold in one case and the absence of a suitable final state in the other prevent the detection of Cosmological Relic antineutrino background unless very specific conditions on the Q-value of the decay are met or significant improvements on the performances of ion storage rings are achieved. The most promising way toward the detection of Cosmological Relic neutrino thus remain the use of $\beta^\pm$ decaying nuclei, as described in [1]. In particular, KATRIN [5] and MARE [6] envisage significant energy resolution improvements in their so called "phase II" toward the direct measurement of the neutrino mass. These improvements, together with an effort to significantly increase the decaying nuclei fiducial mass could really bring to experimental reach the Cosmological Relic Neutrino Background discovery through its direct detection.

References

[1] Cocco A G *et al.* 2007 *Journal of Cosmology and Astroparticle Physics* **06** 015
[2] Bambinek W *et al.* 1977 *Reviews of Modern Physics* **49** 77
[3] Behrens H and Büring W 1982 *Electron Radial Wave Functions and Nuclear Beta Decay* (Clarendon, Oxford)
[4] Cocco A G *et al.* 2009 *to be published*
[5] Drexlin G 2008 *Journal of Physics: Conference Series* **136** 022031
[6] Monfardini A *et al.* 2006 *Nuclear Instruments and Methods in Physics Research A* **559** 346

AUTHOR INDEX

Andriamonje, S 11
Aune, S 11, 85
Ball, M 22
Barata, J A S 22
Barker, G J 72
Borges, F I G M 22
Bougamont, E 11
Calvo, E 22
Cárcel, S 22
Carmona, J M 22
Cebrián, S 22
Cervera, A 22
Chapellier, M 11
Cid, X 22
Cocco, Alfredo G 91
Colas, P 11, 56, 62
Conde, C A N 22
Dafni, T 22, 37, 85
Derré, J 11
Dias, T H V T 22
Díaz, J 22
Dos Santos, J M F 22
Fanourakis, G 85
Fernandes, L M P 22
Ferrer-Ribas, E 11, 22, 37
Freitas, E D C 22
Gaffet, S 11
Galán, J 22
Geralis, T 85
Gerbier, G 11
Giganti, Claudio 5
Gil, A 22
Gil, I 22
Giomataris, G 37
Giomataris, I 11, 22, 56, 62, 85
Gómez, H 22
Gómez-Cadenas, J J 22
Gordetzsky, P 37
Gornea, Razvan 16
Granena, F 22
Grignon, C 56, 62
Gros, M 11
Guillaudin, O 56, 62
Heffner, M 54
Hernando-Morata, J A 22
Iguaz, F J 22, 37, 85
Irastorza, I 11, 22, 22, 37, 85
Koumeir, C 56, 62
Kousouris, K 85
Lacarra, J Galán 85
Lázaro, M 22
Lemrani, R 11
Lightfoot, P K 72
Lopes, J A M 22
Lux, T 22
Luzón, G 22
Mangier, P 11
Martín-Albo, J 22
Martínez, D 22
Massó, Eduard 1
Mavrokoridis, K 72
Mayet, F 56, 62
Mols, J P 85
Monrabal, F 22
Monteiro, C M B 22
Morales, J 22, 85
Muñoz-Vidal, J 22
Navick, X F 11
Nova, F 22
Novella, P 22
Nygren, D 22
Palomares, C 22
Papaevangelou, T 85
Piquemal, P 11
Ramachers, Y A 72
Ribas, E Ferrer 85
Ripoll, L 22
Rodríguez, A 22
Ruz, J 22, 85
Saakyan, Ruben 30
Salin, P 11, 37
Sánchez, N F 22
Santos, D 56, 62
Santos, F P 22
Savvidis, I 11
Sciolla, Gabriella 46
Serra, L 22
Sorel, M 22
Spooner, N J C 72
Távora, L M N 22
Tomás, A 22, 37, 85
Vafeiadis, T 85
Van Der Graaf, Harry 43
Veloso, J F C A 22
Verdugo, A 22
Vergados, J D 11
Vuilleumier, Jean-Luc 67
Yahlali, N 22
Zampaolo, M 11

Institute of Physics
Dirac House, Temple Back
Bristol BS1 6BE UK